MARCONI'S NEW STREET WORKS

1912 – 2012

By the same author

2MT Writtle. The Birth of British Broadcasting   (1988)

Marconi on the Isle of Wight (2000)

2MT Writtle. The Birth of British Broadcasting (2010)

A Kind of Magic
The Birth of the Wireless Age   (2011)

**For more information see
2mtwrittle.com**

MARCONI'S NEW STREET WORKS

1912 – 2012

Birthplace of the Wireless Age

Tim Wander

An Authors OnLine Book

First published in Great Britain 2012

Text Copyright © Tim Wander 2012

Cover Design © Tim Wander 2012
Photographs © GEC-Marconi or where otherwise stated

The Author, Tim Wander, has asserted his moral right to be identified as the author of this work under the Copyright, Designs and Patents Act 1988.

All rights reserved. No part of this publication may be reproduced, stored in a retrieval system, or transmitted in any form or by any means, electronic, mechanical, photocopy, recording or otherwise, without prior written permission of the copyright owner. Nor can it be circulated in any form of binding or cover other than that in which it is published and without similar condition including this condition being imposed on a subsequent purchaser.

www.2mtwrittle.com

ISBN 978-0-7552-0693-3

This book published by:

Authors OnLine Ltd.
19 The Cinques,
Gamlingay, Sandy,
Bedfordshire, SG19 3NU.
England.

A CIP catalogue record for this book is available from the British Library.
Text design and typesetting byTim Winder

This book is also available in e-book format, details of which are available at:
www.authorsonline.co.uk.

'Wireless is now available to shipping for communication by means of long, medium, and short wave telegraphy and telephony, bringing to the ship information regarding time, weather conditions, navigational warnings, and news of happenings in the world at large, as well as facilitating social correspondence for passengers, and keeping ship owners in touch with their ships.

On shore, a system of wireless coast stations has been organised by all maritime countries for communication with ships; and wireless beacon stations have been established for the purpose of providing signals upon which ships can take bearings with their direction finders.

So far as the sea is concerned, therefore, and quite apart from the developments that have taken place in communication on land, I feel that my early belief in the possibility of this form of communication has been fully justified.

As one who is deeply attached to the sea, I am proud to have been able to render this service to the sea-going community.'

Guglielmo Marconi.
On the occasion of the fortieth anniversary of his earliest experiments.

'Radio shall contribute to the progress and common good of people in time of peace. It shall contribute to the triumph of justice and mankind in time of war.'

Guglielmo Marconi to Luigi Solari, just before his death in 1937.

Have I done the world good or have I added a menace?

Guglielmo Marconi.

Marconi's New Street Works.........

I feel like one
　Who treads alone
Some banquet-hall deserted,
　Whose lights are fled,
　Whose garlands dead,
And all but he departed.

Oft in the Stilly Night

Thomas Moore (1779–1852)

For Judith

Contents

Foreword x
Preface xi

Acknowledgements and List of Contributors xiv
Introduction xvii
Photographs and Illustrations xxi
Photographic Credits xxvii
Authors Notes xxix

Chapter 1 Prelude to Wireless 1
Chapter 2 Hall Street - The World's First Wireless Factory 7
Chapter 3 Prelude and Disaster 15
Chapter 4 The World's First Purpose Built Wireless Factory 30
Chapter 5 The War to end all Wars 65
Chapter 6 The Birth of British Broadcasting 78
Chapter 7 Domestic Radios and the BBC 97
Chapter 8 Short Waves 102
Chapter 9 Radar 115
Chapter 10 The Birth of Television 120
Chapter 11 The Second World War 130
Chapter 12 After the War was won 193

Chapter 13 New Ideas	216
Chapter 14 New Street Services	226
Chapter 15 Apprentices	239
Chapter 16 1970 - 2000	263
Chapter 17 Corporate Collapse	301
Chapter 18 The End of a Long Road	308
Chapter 19 After-words in 2012 – One Hundred Years	317
Appendix 1 A Hundred Years Of Change - Maps and Photographs	333
Appendix 2 The Marconi Company 1937 – 1999	355
Bibliography	358
Index	359
About the Author	366

Guglielmo Marconi

FOREWORD

Marconi's New Street Works
The Saga Continues......

The saga of our beloved factory continues. The buildings are falling into disrepair, the windows of the New Street frontage are boarded up, the garden has not been tended for at least four years and buddleia some eight feet high is growing out of the flower beds and some of the walls.

Various attempts have been made to get Chelmsford Borough Council to tidy the front area but to no avail. The administrators have been approached following break-ins and their only comment is that they are trying to sell the property and meanwhile have put in some CCTV to prevent vandalism. There is however little evidence of the presence of CCTV.

According to the local weekly newspapers, one of the big supermarket chains has expressed an interest in purchasing the site but nothing has been officially announced. It is alleged that a certain group is thinking about purchasing the building facing New Street and turning this into a museum and art centre. Again, nothing official has been released.

2012 sees the centenary of this building. It is hoped that some firm plans for its occupation and usage will be announced before the Marconi Veterans' Association celebrates this anniversary.

Many people have commented on how Marconi has been neglected in Chelmsford and the wonderful legacy he left. It would appear the leading citizens of Chelmsford are not interested in this or any of the legacies left by other great industrialists who set up businesses in the town such as Hoffmann, Crompton and Christy. It could become a major tourist attraction if only the local Chelmsford Council attached much more importance to the heritage bequeathed by these pioneers.

Peter Turrall MBE

PREFACE

The essence of the Marconi Company approach to business was always centred on its people, starting with Guglielmo Marconi and the team of world class engineers he first gathered round him. Once upon a time, to be known as a 'Marconi man or woman' was to be acknowledged as belonging to a rather special organisation.

In the book 'Marconi 1939-1945 - A War Record', the author wrote that a Marconi research engineer, walking in the soft sunshine on the little lawn beside Marconi College was looking back over the feverish war years was trying to explain the scale of scientific and manufacturing work undertaken by the Company. He paused, considered for a moment and then said, as if stating an objective scientific fact, that he felt that the Marconi Company *had a soul.*

Generations of families, many of whom passed through apprentice training and service personnel who joined in the post-war period all contributed to this reputation in their various ways.

> '**Cut** us anywhere and I am sure we would have Marconi written all the way through us like a bit of rock.'
>
> *Derek Pope*

> '**Well** it was a family, we had our little Marconi badge which we put on and we were very proud of it.... Most of those were badges that really should have been on equipment.... they had a nut and bolt, a nut on the back and you used to put them on your coat. We were very proud of it and I reckoned if you went to any seaside place, I mean in those days people didn't go abroad for their holidays it didn't matter where you went for our seaside holiday you invariably bumped into the Marconi man, who was always there in person, and also, on two

occasions I went to hospital there was the Marconi person in the ward . . . you felt comfortable because they was all Marconi people and you felt part of a family, you know.

David Newman and Derek Pope

'**How** did we Marconi engineers ever have time for both Firm and Family?'

E.G. Walker

From the first days of this project, it was clear that in just a few hundred pages I could never hope to document a hundred years of world beating technologies, nor attempt to give justice to any discussion about the huge industrial and social changes that the last century had witnessed.

More importantly, it was clear that a dry list of projects, buildings and equipment, even if mixed with some of the Company's successes and occasional failures would render the text unreadable.

For me the key part of the New Street story surrounds the people who worked there and often gave their entire working lives to the Company. But again how do you even touch upon recording the careers and actions of the hundreds of thousands of people who have stepped through the doors of Marconi's New Street Works?

But the fast approaching Centenary demanded that I try.

What follows is a tribute to all that happened at the works and all the people who served there. It tells a small part of the story of the world's first purpose built wireless factory, a hundred years of innovation and invention, challenges and change. It is the story of the rise and fall of a great Company and in part it is the story of the man who brought the science and art of wireless communication to the world.

Part history and part oral record, here then are just some of the stories, memories, reminiscences, anecdotes and thoughts from the people who were all part of the

New Street story over the past one hundred years. Because they are personal accounts, not cold hard histories, the order in which they appear may sometimes seem imperfect, but in careers sometimes spanning over 45 years chronology is often not the most important thing. Recalling experiences sometimes 70 years after the event can also lead to recollections that are understandably not accurate in every single detail.

But what cannot be taken away are the feelings, interest, passion and sometimes great sadness of the memories. It was been a privilege to speak to and write the words of people who were forging careers and working and fighting for their country long before I was born.

These then are their words. The Marconi men and women. I have only been the catalyst that put them all together and provided the glue to join them to the page.

Tim Wander
October 2011

Acknowledgements
and
List of Contributors

Over the past two years many, many people have taken the time to E-mail, write or speak to me. Some simply offered their best wishes and support, others memories and information about The Marconi Company and their time there, even if they were not really part of the New Street story. All were gratefully received and documented. Interestingly, in all the comments and thoughts I never found any sense of criticism or regret, unless the subject of the current state of the Works was mentioned.

I take this opportunity to thank you all for your support for the project. I have lived with the story of the birth of wireless and radio broadcasting for nearly thirty years now, but I was genuinely surprised by the depth of feeling and even emotion that the once great factory still stirs in people who are now often long past retirement age. I hope I have done this some small justice.

I must single out a few people for thanks. As always my long suffering wife Judith who has transcribed every handwritten note and letter received along with all my scrawled sheets, saying nothing when whole pages of typing were rejected and deleted at a single stroke. To Dr Geoff Bowles, proof reader extraordinaire who yet again attacked the text and my love of far too many words in a small sentence with great dedication.

Many thanks to Peter Turrall MBE from the Marconi Veterans Association who has long campaigned for better recognition of Marconi's work in Chelmsford and who supported this project from the start. He also volunteered from the outset to proofread the first draft – not an easy task. Of course any errors that crept through are still all my own work.

The following is a list of Contributors whose words I have included, perhaps just one line or sometimes a hundred and one, but I thank you all.

If I haven't included everything you sent, or your words didn't make it past the editors blue pen (that shows my age), then I can only apologise and claim pressure of space.

If I have made a mistake I can only claim old age and failing memory.

Much like us all then.

Those who remembered the Marconi New Street Works included:

Alice M Atkinson	*Stuart Axe*	*Bill Barbone OBE*
Norman Bartlett	*P. Betts*	*Diane Berthelot (nee Lawson)*
Ted Bocking	*A.J. Brown*	*Alan Buckroyd*
Eileen Carter (nee Evans)	*Stan Church*	*Gerald Cock*
Joan Cockerton	*Doris Colliston*	*Olive Cox*
Vera Crause	*John Crouch*	*Mrs Dilys Edwards*
Peter Eckersley	*Ray Fautley*	*Charles Franklin*
C Frost	*Martin Gebel*	*Lilian Gowers (nee Baldwin)*
Joan Griggs	*Tom Gutteridge*	*Eileen Hance*
Brian Hanchett	*Peter Helsdon*	*Gerald Hockley*
Roger Hunt	*Kenneth Hutley*	*Phil Hollington*
Roy Hubbard	*Margaret Hudgell*	*Eunice M. Jackson*
David A. Johnson	*Ronald Kitchen*	*M. Knight*
James C.H. Macbeth	*Roderick Mackley*	*George Maclean*
Guglielmo Marconi	*Dave Mckay*	*Bill Meeham*
Dame Nellie Melba	*Betty Morris*	*Caroline Moth*
Geoffrey Nash	*David Newman*	*E.H Palmer*
Eric Peachey	*Nick Pettefer*	*Derek Pope*
Catriona Potter	*Richard (Rick) Potter*	*Steve Readings*
Frederick Roberts	*Keith Ronaldson*	*Henry Joseph Round MC*
Hilda Rounce	*David Samways*	*Winifred Sayer*
Isaac Shoenberg	*Hector Norman Smith*	*Barbara M. Stephens*
Summers (Mrs)	*Ed Swain*	*Dawn Swindells*
Keith Thomas	*Brian Thwaite*	*Peter Turrall MBE*
Bertie Upston	*Steve Wakerley*	*E G. Walker*
Judith Wander (nee Paskins)	*Tim Wander*	*George Warner*
Doris Warren	*Frank Whybrow*	*Joan Wigley*
Chris Wild	*Len Wilkinson*	*Stephen Williams*
Susan Wilson	*Sam Woollard*	*Peter Wright*

Thank you all........

INTRODUCTION

The Marconi New Street Works in Chelmsford Essex was the world's first purpose built wireless factory and became the world's first electronics factory to use mass production techniques. For well over ninety years the huge factory was the manufacturing centre of the massive Marconi Company that stretched across the world. It's success and growth gave Chelmsford the worthy title of 'Home of Radio' although perhaps a better title would be the 'Home of Wireless', for this is where wireless was born and the radio age was built.

However, the disastrous collapse of The Marconi Company in the first part of the 21st Century prompted the eventual abandonment of the huge complex of offices, workshops, laboratories, test areas and manufacturing lines that in his heyday once employed well over 10,000 people. Today, in his hundreth year, the site currently stands empty and derelict. It has been extensively vandalised, its lead roof flashing and fixtures stripped out and stolen, whilst the gardens at the front and even the main yard are rapidly becoming overgrown.

Part of the historic Marconi New Street site had already been lost to the new Eastwood House development started in 1992. During the construction of the new Marconi Radar offices at the top of the New Street site the historic blacksmiths forges, many buildings (including building 46), air raid shelters, workshops, one of the original massive concrete aerial mast bases and even the room where Dame Nellie Melba gave her historic broadcast were swept away.

The New Street factory opened in 1912, the year the SS *Titanic* was lost in the middle of the Atlantic. The 711 lives that were saved were solely due to the use of Guglielmo Marconi's invention and the brave Marconi Wireless Operators aboard, one of whom gave his life.

The world of 1912 was a very different place. It is hard to imagine now, but one hundred years ago everything we now take for granted was all an unthinkable dream.

In 1912 motor cars were still a novelty and mechanised flight was in its infancy, largely the preserve of the military and air shows. Even the Royal Flying Corps, predecessor of the RAF was only formed in 1912. The very thought of commercial air travel and the ability to travel across the world in less than half a day was the stuff of science fiction writers. Even they never dreamt that it would be possible to have the ability to communicate with each other anywhere on the face of the planet or to be able to access a wealth of information and entertainment at our finger tips in an instant via a world wide computer network.

In the world of 1912 Queen Victoria had only died eleven years previously, her grandson George V was on the throne and Herbert Asquith was the Liberal Prime Minister. Women could only vote if they owned property and strikes and riots featured often during this period with coal miners and Dockers taking to the streets.

The shadow of global war was also being cast across Europe with the outbreak of war in the Balkans and there were growing concerns about the political situation in many European countries. There was also a rapidly escalating arms race fuelled by the German Empires' plan to create an Atlantic fleet that could challenge British naval power.

During the First World War, much of the wireless equipment used by the Army Navy and Royal Flying Corps was built at New Street. The critical radio direction finding equipment for the Royal Navy at the battle of Jutland and elsewhere were designed and built there, as were the air traffic control systems, radio direction finding systems and the airborne wireless equipment which supported the birth and rapid growth of international air traffic.

After the war, Chelmsford hosted this country's first wireless news reports and music concerts that within two years gave us British broadcasting and the birth of the BBC. Britain's first fledging television service was developed there in the 1930s. During the Second World War, although designed at the nearby Writtle site many of the aircraft transmitters (T1154) and receivers (R1155) for Bomber and Coastal Command were also built at New Street. The Royal Navy's crucial

CR100 receiver was part built and tested there as were a large number of 'spy' sets dropped in occupied Europe with the SOE. The factory then went on to give birth to the advent of a national television broadcasting service as well as radar, the first advanced mainframe computers, and generations of the world's best radio and television transmitters and receivers.

At New Street, Marconi Communication Systems produced some of the finest communication equipment, television and sound broadcasting transmitters, black and white and colour television cameras, telecine, outside broadcast vehicles, mobile radios, satellite communication and marine equipment.

Marconi Radar Systems was formed in New Street and then moved to the old Crompton Parkinson factory in Writtle Road Chelmsford. The Company went on to produce some of the world's best radar equipment, extensively used in airports, civil and military sites as well as on ships and aircraft.

The Marconi Aeronautical Division was formed in New Street and moved in the 1950s to new premises in Basildon to manufacture products for both civil and military applications. The Marconi Research establishment was formed at New Street (and Broomfield) and moved to Great Baddow just outside Chelmsford, providing many of the development requirements and advanced techniques for products utilised in all the Marconi Companies. Marconi Research established hundreds of patents and was part of the development of many world leading technologies, not least of which was the invention of the liquid crystal display and of course radar.

Marconi Marine also originated from the New Street Works and moved to set up a factory and offices at Westway in Chelmsford, where they produced wireless and navigation equipment for ships and ports as well as training wireless operators engaged by many shipping companies throughout the world.

Marconi Communications Systems at New Street pioneered the development of the huge Kilostream network for BT, satellite communications and a host of other lesser, but none the less cutting edge projects.

Overall it is not a bad record.

I believe that the New Street Works is one of this country's, if the not the world's, most important sites of industrial heritage and archaeology. But only a fraction of the facade and buildings are protected by listed status and the rest is now faced with imminent demolition. Like many other industrial buildings and areas that gave birth to the modern age, it may well be lost forever beneath a housing estate or shopping centre.

This then is the story of the world's first wireless factory, part told in the words of the people who worked there. Sometimes many generations of the same family gave their service and loyalty to the Company but they have all left now, having watched the once proud factory turn into a disused shell. Only their memories remain.

Once the New Street Works had over 10,000 people working each day inside its walls, but the site closed its doors for the last time in 2008, having been swamped by the shadow of the massive mismanagement and corporate incompetence that caused the collapse of the once proud Marconi Company.

Without any doubt, within those factory walls, the science and art of wireless communication was started. In fact I firmly believe that this is also where the modern electronics age, be it radio, television, radar, satellite or even mobile telephones was born.

Britain lived and prospered through the revolutions bought by the age of steam and the age of steel. Marconi's New Street Works from 1912 to 2012 was and still is the birthplace of Britain's last industrial revolution.

The age of wireless.

Photographs and Illustrations

Cover

New Street Works, 1920 *(MWT)*
Dame Nellie Melba, 1922 *(MWT)*
New Street Works, main manufacturing, 1916 *(MWT)*

Back Cover

New Street Works and Goods Yard, 1920 *(MWT)*
New Street Works, 1944 *(MWT)*
New Street Works, 2010

Chapter One

1. **Guglielmo Marconi** *(MWT)*
2. **Victoria Street**

Chapter Two

3. **The Marconi Hall Street Works in Chelmsford** *(MWT)*
4. **Hall Street Works interior** *(MWT)*
5. **Dalston Works** *(MWT)*
6. **Dalston Works interior** *(MWT)*
7. **Punch Cartoon for Mr Marconi**

Chapter Four

8. **Breaking Ground Ceremony** *(MWT)*
9. **Breaking Ground Ceremony – Marconi** *(MWT)*

10. **Foundations** *(MWT)*
11. **Part Built** *(MWT)*
12. **New Street Construction Team** *(MWT)*
13. **Scaffolding to main building** *(MWT)*
14. **Finishing Off** *(MWT)*
15. **New Street Annotated Department Map and Key** *(MWT)*
16. **Machine Shop** *(MWT)*
17. **Mounting Shop** *(MWT)*
18. **Winding Shop** *(MWT)*
19. **Winding Shop** *(MWT)*
20. **Paint Shop** *(MWT)*
21. **Raw Stores** *(MWT)*
22. **Instrument Test** *(MWT)*
23. **Instrument Test** *(MWT)*
24. **Accounts** *(MWT)*
25. **Drawing Office** *(MWT)*
26. **Case Construction**
27. **Packing Shop**
28. **Examiners Department** *(MWT)*
29. **Assembly Shop** *(MWT)*
30. **Finished Parts Stores** *(MWT)*
31. **Finished Stores** *(MWT)*
32. **Surgery** *(MWT)*
33. **Blacksmiths Shop** *(MWT)*
34. **Power Test** *(MWT)*
35. **Power Test** *(MWT)*
36. **Power Test** *(MWT)*
37. **New Street c 1915** *(MWT)*
38. **Show Room, 15kW Transmitter** *(MWT)*
39. **Show Room** *(MWT)*
40. **Aerial Leadin c 1933**
41. **Original 1912, 250 ft Mast** *(MWT)*
42. **New mast Construction** *(MWT)*
43. **New Street Masts** *(MWT)*
44. **New Street Masts**
45. **New Street Masts**

46. **New Street Masts**
47. **New Street Masts** *(MWT)*
48. **Mast Demolition, 1935** *(MWT)*
49. **Demolished mast section** *(MWT)*

Chapter Six

50. **W.T Ditcham** *(MWT)*
51. **Dame Nellie Melba** *(MWT)*
52. **Lauritz Melchior** *(MWT)*
53. **Writtle, site of 2MT** *(MWT)*

Chapter Eight

54. **SWB 1 Transmitter**
55. **SWB 8 Transmitter**
56. **SWB 11 Transmitter**
57. **Daventry 5XX Transmitter**
58. **One of the first PBX telephone systems, New Street, 1929.**
59. **Sports Day, 1932**

Chapter Ten

60. **Alexander Palace Television Transmitter**
61. **Outside Broadcast Vehicles, 1952**

Chapter Eleven

62. **Marconi House in Camouflage** *(MWT)*
63. **Marconi Outing, Hastings, June 1955**
64. **Marconi Bomb Disposal Group**
65. **Marconi New Street Camouflage Seventy years on**
66. **Marconi New Street Camouflage Seventy years on**
67. **Marconi New Street Camouflage Seventy years on**

Chapter Twelve

68. Machine Filing Department 1956
69. New Street Masts
70. New Street Masts
71. Mast Bolt
72. Mast Bolt

Chapter Fourteen

73. Marconi Fire Brigade
74. Marconi Fire Engine

Chapter Sixteen

75. Marconi New Street Manufacturing, 1980's *(MWT)*
76. UKMACCS – Part of the ARC Team
77. UKMACCS ICS 3 Power Bank
78. UKMACCS Equipment rack - RPI and QCI
79. 10kW Amplifier
80. Judith and the VAX Computer Room

Chapter Eighteen

81. Hall Street, 2011

Chapter Nineteen

82. Front View New Street, 2011
83. Front View New Street, 2011
84. Blue Plaque
85. Front Gate, 2011
86. Main Yard, 2011
87. Main Yard, 2011
88. Main Yard, 2011
89. Goods Yard, 2011

90. Equipment destined for the scrapheap
91. The End of an Era
92. The Last Lightswitches
93. New Street main entrance stairs in beter times
94. Marconi House foyer in better times
95. Main Yard looking toward New Street
96. Building 720 Roof
97. Building 720 groundfloor Canteen
98. Main Works shrapnel damage to beam from German Bomb
99. Main Works shrapnel damage to wall
100. Goods Yard
101. Goods Yard
102. Manufacturing, July 2011
103. Manufacturing , July 2011
104. My old area – CSU top floor Building 720, July 2011

Appendix One – New Street through the Decades

105. **1912 Plan**
106. **Front view, c1915**
107. **Front view, c1915**
108. **1921 Plan**
109. **Aerial Photograph, c1920** *(MWT)*
110. **Aerial Photograph, c1920** *(MWT)*
111. **1927 Plan**
112. **Marconi New Street Frontage, c1930**
113. **Marconi New Street Yard, c1930** *(MWT)*
114. **1936 Plan**
115. **1938 Plan**
116. **1941 Plan**
117. **New Street, c1943, Luftwaffe Aerial target map**
118. **Building 720 under construction**
119. **Marconi New Street Yard, c1950** *(MWT)*
120. **Marconi New Street Frontage, c1950** *(MWT)*
121. **Aerial Photograph, c1955** *(MWT)*
122. **Aerial Photograph, c1955** *(MWT)*

123. **Building 46**
124. **Building 46 Air Raid Shelter**
125. **1969 Plan**
126. **Looking toward Building 46 from 720** *(MWT)*
127. **Marconi New Street Goods Yard and Water Tower** *(MWT)*
128. **Marconi's New Street Yard, c1984**
129. **1980 Plan**
130. **2000 Plan**
131. **Construction Eastwood House, c1992**
132. **Aerial view, Construction Eastwood House, c1992**
133. **The Author, New Street, 2012**

Photo Credits

All photographs marked *(MWT)* are reproduced by kind permission of what I knew as GEC-Marconi Ltd. In this text I have, for the sake of brevity and clarity usually simply referred to the Company as 'Marconi's'.

A number of photographs in this book have been published for the first time, especially in Chapter Four where some of the images of the Works were saved by the author (and David in the Marconi photographic department) from the original glass plate negatives that had already broken up.

Marconi's was founded by Guglielmo Marconi in 1897 as the Wireless Telegraph and Signal Company Ltd. It was renamed Marconi's Wireless Telegraph Company Ltd. in 1900 and The Marconi Company (Ltd.) in 1963. After this it had various names due to its ownership by GEC and later by others. A more complete list is shown in Appendix Two.

Other photographs are © Authors Collection. Every effort has been made to fulfil requirements for reproducing copyright material although most of the images are now at least 100 years old. Some images (or other media files) are now in the public domain because their copyright has expired. This applies to Australia, the European Union and those countries with a copyright term of life of the author plus 70 years. The author would be glad to rectify any omissions at the earliest opportunity.

The stories and the personal photographs in this book have been contributed by the public and copyright for both rests with the authors, although by contributing to this book they granted the author a non-exclusive right to sublicense and use the content.

I would like to thank Neil and Nelly who managed to record the state of the Marconi New Street Works interior in 2011 and allowed me to use some of their photographs in Chapter 19.

Also my thanks to Martin Bates who had developed an excellent Marconi website and spent hours producing the plans of the New Street works.

Some of the memories in this book have also been submitted, sometimes in more detailed forms, to other archives including the Essex Museum Service and WW2 People's War, an online archive of wartime memories contributed by members of the public and gathered by the BBC. The full archive can be found at bbc.co.uk/ww2peopleswar.

Part of this text was abridged from *2MT Writtle, The Birth of British Broadcasting* by Tim Wander published in 2010. ISBN 978-07552-0607-0.

For more information see 2mtwrittle.com.

Authors Notes

Essentially the terms *wireless* and *radio* can mean exactly the same thing. Surprisingly it was the word radio that first came into use even before Heinrich Hertz's proved the existence of electromagnetic waves, but despite this very early usage, it has always been regarded as the modern version of the term wireless.

Although both terms are correct, the Americans adopted the term radio in common usage earlier than the UK. It was used by American pioneer Lee De Forest as early as 1907 and formally adopted by the US Navy in 1912. On 28th September 1923 the BBC officially accepted 'Radio' by publishing the 'Radio Times.'

The word 'wireless', when used in relation to broadcasting or radio equipment is now commonly regarded as an old fashioned term. However, in recent years the term 'wireless' has gained renewed popularity through the rapid growth of short-range computer networking such as Wireless Local Area Networks (WLAN), Bluetooth and Wi-Fi

Also I was recently reminded that a young Italian called Guglielmo Marconi once called his Company the Marconi *Wireless* Telegraph Company Ltd, and how could he be wrong?

Tim Wander
September 2011.

In the end, when you have finished the book, if you have any comments, thoughts, corrections, omissions or would even like to add your reminiscences, then there is a chance.

In 2013, due to the wonders of print on demand this book can be reviewed, corrected, updated and new photographs and text can be added. Please email timwander@compuserve.com.

CHAPTER ONE

Prelude to Wireless

It is February 1896. Two lone figures stand on the platform of London's Victoria station; a man and a woman. They are mother and son, well dressed and surrounded by luggage and travelling trunks, but the man is wearing an unusual deer stalker hat and has a thick fur coat pulled tight around his thin frame. By his feet he also has two rather incongruous black boxes. He is waiting for his mother's cousin to collect them and hopefully introduce him and his new idea into the scientific community of late Victorian England.

The young man, half Italian and half Irish is Guglielmo Marconi and he has a dream of changing the world with a communication system that does not need wires.

The pair have already travelled a long way across Europe from Italy. They are tired, cold and frustrated at having been delayed in British customs for many hours. The hour is late and they still have a long way to go. Waiting in the shadow of the great Grosvenor Hotel, the Victoria Station yard is one of the great terminal stations for the omnibuses from all quarters of the Metropolis. This, combined with the busy horse drawn Hansom cab traffic could not have failed to impress them. Any visitor from the Continent would soon come to sense the bewildering energy of the mighty city. As Marconi caught his first glimpses of the unfamiliar capital, the young man, barely 21 years old, could never have dreamt how far this journey would take him.

His cousin, Henry Jameson-Davis arrived. A first cousin, but he hasn't met the young man since he was six years old. Little did he know that his journey across London in the horse drawn Hansom cab to collect the visitors would also change his life forever.

Victoria Street Station, c1896

As the new century approached, Victorian England toasted the advent of a new scientific age. Between 1885 and 1889 the German physicist Heinrich Rudolf Hertz had demonstrated that the Scottish physicist and mathematician James Clerk Maxwell's theoretical electromagnetic waves were a fact. He could generate them, receive them and even measure their properties and they quickly became known as 'Hertzian Waves' or even 'wireless waves.' Hertz was one of the best scientists of his generation, but he died in 1892 at the very young age of 36 and never saw any practical use for his laboratory experiment.

Heinrich's 'Hertzian waves' or 'aetheric waves', were still little more than a scientific curiosity when a number of other inventors and scientists, including Professor Oliver Lodge, started organising lectures and practical demonstrations repeating his work. But like many other early pioneers even Lodge failed to see any commercial potential in his experimental results.

Part of the problem was that the prominent scientists of the time had announced that the curvature of the earth would severely limit the range of these new waves and consequently they had no practical purpose. It was also believed that the powerful waves from any high powered long distance station would swamp the feebler waves from ordinary ship and shore stations, producing nothing but chaos in any receiver. The established scientific community was adamant that wireless communication by electromagnetic waves had no future and could never challenge the well-developed cable telegraph system and network.

In 1896 the young Italian inventor arrived on Britain's shores with a dream of transmitting messages in all weathers and at any time and place, without the use of wires. He had just two boxes filled with some barely working and rather crude wireless equipment, fresh from his experimental work bench that had already been damaged by wary customs officers.

He had an introduction to a first cousin that he had not seen since he was a child, but he had no staff, no finance, no offices or workshop, no scientific training, formal education and perhaps most importantly for Victorian England, no reputation. What he did have was an enormous passion and an almost unbelievable belief in himself, his ideas and his system.

He set out on a determined, almost obsessive path to prove that his laboratory experiments were the basis of a reliable, long distance communication system. From the moment he first set foot in England, Marconi's public demonstrations and private tests were each designed to move his system one step further, both in technological development and in commercial understanding and acceptance.

The problem for Marconi was that the great Heinrich Hertz had shown that his invisible waves obeyed exactly the same laws of reflection and refraction as did light waves, and that in fact the only fundamental difference between the two was one of frequency. These conclusions had also been verified time and again by Marconi and other scientific workers. Yet equally beyond dispute, was the fact that the data being amassed from Marconi's demonstrations and experiments at numerous wireless stations around the world showed a steady upward progress in the ranges that could be reliably achieved. For some reason Hertzian waves were breaking the 'straight line' rule. At first the difference between theory and observation was small, but it soon rose to double the predicted figure. When distances of eighty miles could be guaranteed, and Marconi could reliably communicate with ships and other stations far below the horizon this anomaly could no longer be ignored. Still the common scientific opinion of the day simply chose to doubt the authenticity of these experimental results.

At times it was only the charisma of the young Italian inventor that kept interest alive. Guglielmo Marconi was convinced that these eminent men were wrong and that reliable long distance communication was just a matter of increasing transmitted powers and improving the sensitivity of the receivers. This belief was soon backed by his repeated observations of successful transmissions to ships lying below the horizon, suggesting that the propagation of wireless waves was far more complex than the 'straight line' or 'line of sight' theorists believed.

For the first five years after arriving on England's shores, Guglielmo Marconi had two driving ambitions that fuelled his research. He was determined to quash the persistent rumours regarding the accuracy of the ranges he claimed for his system. A practical and reliable range of at least 25 miles in all weathers was essential to his dream of providing systems to safeguard ships at sea. By the end of 1897 his system could do that and more every day and in every type of weather.

It is easy to forget that before Marconi and the widespread introduction of his reliable wireless communication system, any ship, large or small, commercial, civilian or military effectively disappeared as soon as it left sight of land. Sometimes it never returned and the owners and relatives of the passengers and crew would never know their fate.

Marconi's second goal was perhaps more ambitious. He was determined to set his system in direct competition with the long distance undersea telegraph cable companies. Up until 1899 wireless communication was still largely retracing the path blazed by the cable telegraph half a century earlier, developing point to point communication, albeit now without the need for expensive connecting wires.

But developments moved rapidly. As the new century approached there were the beginnings of talk about innovations which moved beyond what the cable telegraph could do. People started to understand that the new 'wireless' could be useful in aiding safety at sea. Marconi also saw a role for the new system in 'the warfare of the future', and also the potential to someday compete with the telephone in providing personal communication.

By 1899 Marconi had already built the world's first permanent wireless station at Alum Bay on the Isle of Wight and new stations had been established on lightships and on the English east and south coasts. Marconi's wireless system instigated the first ship to shore message, the first shipwreck rescue, first use of the international distress signal and undertook the first transmissions to France across the English Channel. He established the first use of wireless in the air and after extensive trials with the Royal Navy he sent his fragile new system to face the harsh environment and demands of the military during war for the first time in South Africa.

Throughout this period, Marconi was determined to maintain the technological integrity of his Company, never publicising new inventions until they had been fully proved or making promises which he could not substantiate. Through it all the young Italian displayed a considerable flair for publicity and a deft hand with the press, public, military authorities and even the doubting scientific community.

Being Marconi, he tackled all these head on with a quiet reserve but ruthless dedication to the task as he moved from experiment to demonstration and from test to trials, each carefully calculated step pushing his system, himself and his team harder and harder. Young as Marconi was, his dedication and single-mindedness coupled with his gentlemanly demeanour was very different from the popular Victorian image of the 'mad inventor', and completely different from some of his more illustrious competitors. His continuous successes despite the many obstacles, inspired loyalty in his small workforce of engineers, most of whom had learned their trade in the established Victorian business of telegraph cables. Guglielmo Marconi once said, with some modesty:

> 'I have striven to give the world improved and cheaper means of communication by means of electrical transmission through space.'

Marconi patented his system in June 1896 with the first ever true wireless patent when he was just 22 years old. His Company, formed in 1897 to use his patents, was called the Wireless Telegraph and Signal Co Ltd. This was changed to Marconi's Wireless Telegraph Co Ltd. (often known by its initials MWTCo) in 1901. Despite many competitors it was his Company that set out to develop the science of wireless communication.

Of course such advances would needed talented men of vision and passion who were prepared and able to take on and meet the challenge. Marconi built around him a team that included some of the best engineers in the world, drawing people from many disciplines to work at his side as he built his Company.

But out of all the early pioneers and scientists it was Marconi alone who had the personal drive and single minded belief to develop reliable wireless communication against enormous odds.

CHAPTER TWO

HALL STREET

The World's First Wireless Factory

By the end of 1898 Marconi has spent years three hectic years testing, trialling and demonstrating his system for the Royal Navy, British Army, United States Navy (and Army), Trinity House, the British Royal family and newspapers across the world. He had proved his case. His wireless telegraphy system was no longer a simple laboratory experiment.

To meet the anticipated demand for new equipment Marconi had to urgently seek new premises for manufacturing and administration. The Company's existing Head Office at 28 Mark Lane in the City of London was already overcrowded and could never support the proposed expansion or any form of large scale manufacturing. Marconi's (and the world's) first wireless station at Alum Bay on the Isle of Wight was based in a rented Hotel. The new Haven Hotel wireless station near Poole on the South coast of England was also based in leased rooms. It had only been in operation for two months and its limited space was earmarked solely for research and development. Marconi made the decision to move very rapidly, the choice being a large building located in Hall Street in the town of Chelmsford in Essex.

The reason why Marconi chose Chelmsford is unclear. Two key issues must have been that buildings were far cheaper and larger outside London and the County of Essex is very flat, ideal for wireless experiments and erection of masts and aerials. Chelmsford also had a direct rail link into the capital and was reasonably near the Port of London whose huge volume of shipping represented one of Marconi's immediate potential market places.

In London Marconi enjoyed the social life and understood that it was the centre for finance, science and government, but the combination of sewage, coal fires and unwashed bodies meant that in reality the odour of London was horrendous. Both the rich and the poor had to contend with the evil air around the city. Every surface was coated with soot from the use of coal. New buildings being constructed of Portland stone didn't stay pristine for long. The air people breathed was often foggy and tainted with smoke. For Marconi, more used to England's barren headlands and the rolling hills of Northern Italy, Chelmsford may well have been a welcome respite.

Chelmsford already had a long tradition of industry. Originally local processing and service trades dominated, including flour milling, breweries, tanneries, rope making and iron works. But the opening of the Chelmer and Blackwater Navigation Canal in 1797 encouraged industrial growth as Chelmsford now had a direct trade link to the sea. When the railway finally connected London to Chelmsford in 1843, the complete infrastructure was in place to support a growing industrial town. The railways moved goods, food and people faster than canals or horse drawn wagons and had become the greatest factor in transforming Britain into an industrial nation.

From Marconi's point of view Chelmsford was ideal. He was still near to London, the financial centre of England and indeed London would remain the headquarters of the Company until the day it closed over a century later. But at just over 30 miles from the centre of London, Chelmsford was conveniently outside the area in which the General Post Office had an absolute monopoly over all Telegraphic Communications. There was also plenty of electrical power available as Crompton Parkinson Ltd already had power stations in the town, yet the town was still free of much of the electrical interference from tramways and lifts that had often plagued his wireless trials and experiments in the City. He also found that with plenty of manufacturing works, including Crompton's already established in the town, he had a ready supply of skilled workers.

In the end the choice was probably influenced by a combination of all these factors, or perhaps it was just that suitable premises in Hall Street became available at the right time. Marconi promptly leased Messrs Wenleys' furniture store in Hall Street. When it was built by John Hall in 1858, it was a state of the art steam driven silk mill. John Hall had worked the mill until 1861 when the Government repealed the silk duties, and French imports ruined the Essex silk industry. The Hall Street mill closed in 1863, but Samuel Courtaulds of Braintree, who survived the disaster, ran the mill from 1865 until 1892. It then became Messrs Wenleys' furniture storage depot. The site was ideal for Marconi's requirements especially as the engine house still existed as did the line shafts to power equipment inside the works. The site also had a house attached, now called Alfred Cottage, (possibly after Marconi's brother Alfredo) which provided accommodation or additional office space.

Hall Street now became the world's first wireless equipment factory. At first the site employed just 26 men and 2 boys. They fashioned spark wireless sets and coherer receivers, eventually for the Royal Navy, Lloyds, numerous merchant ships and a generation of great Ocean liners. Hall Street would also make many of the parts for the high-power station at Poldhu Point in Cornwall.

The Marconi Hall Street Works in Chelmsford

Hall Street Works Interior

Wireless was still in its infancy. Initially the Company struggled for income, so it had to diversify into manufacturing motor-car ignition coils, X-ray apparatus and other scientific equipment. Until the Hall Street Works came into operation, any wireless equipment that had been built had been constructed as required, using various modified apparatus bought from established scientific laboratory suppliers. Marconi, his right hand man Gorge Kemp and several assistants hand constructed other specialist parts, but they could never hope to cope with quantity production of commercial equipment.

It must be remembered that at the turn of the century Marconi's was still a small organisation developing an unknown field. During these early years any new engineer considered himself to be part of an elite group, learning the new science of wireless communication as it was invented, perhaps even working alongside Marconi himself.

The formation of the Marconi Hall Street Works now put the whole manufacturing system on a much more formal level, with new departments responsible solely for their own specific areas of research, design and manufacture. The condenser and winding shop, mounting and machining shops all found their way into the new factory moulded into an organised commercial concern under the personal supervision of the new Works Manager, Mr E.T. Priddle. In September 1899 a transmitting station was established on the other side of Hall Street to test equipment as it came off the production line and the Hall Street mast soon became one of Chelmsford's landmarks.

The Hall Street station along with a new station constructed at Dovercourt near Harwich, brought the total number of wireless stations built by Marconi since December 1897 to twelve, including Niton (call sign NI), Haven Hotel (HH), Lizard (LD), Poldhu (CC), Chelmsford (CD), Holyhead (HD), Caister-on-Sea (CS), North Foreland (NF), Withernsea (WS), Rosslare (Ireland RL), Port Steward (from 1902 Malin Head, Ireland MH) and Crookhaven (from 1902 moved to Browhead CK).

On 24th March 1900 the Wireless Telegraph and Signal Company Ltd was reconstituted to become Marconi's Wireless Telegraph Company Ltd. The name change was a Board decision, despite Marconi's objection. The Company considered that given Marconi's international fame, his very name was now a major asset. Even in the days before tabloid newspapers and paparazzi, the pressures of success were building on the young Italian's shoulders. Until this time, Marconi had been personally consulted upon all enquiries and decisions; he then gave his instructions and Managing Director Henry Jameson-Davis delegated members of the staff to deal with them. But the Marconi's Company was now growing rapidly and it was becoming impossible for Senor Marconi to be everywhere and involved in everything.

With the formation of the new Company Jameson-Davis stood down (as he had always intended to do) and a new Managing Director, Samuel Flood-Page had to take the helm. He found himself in technical control of the growing Chelmsford Hall Street manufacturing works as well as having to deal with the increasing spate of technical inquiries. To share the load somewhat, Flood-Page appointed a Works Manager, Mr H. Cuthbert Hall, who took the post early in 1901.

At the same time, Mr E.A.N. Pochin, a qualified industrial electrical engineer, assumed responsibility for the technical control of the Company.

In late January 1901, HMS *Jaseur* using Marconi's new tuned receiver successfully received signals from both the Haven Hotel wireless station and from HMS *Hector*. This is the first time at sea that two stations were received by a single vessel with one receiver. In 1901 a further 50 'service sets' were ordered for fleet use. All these sets were manufactured at the Marconi Hall Street Works in Chelmsford. By 31st December 1901 the Royal Navy had 105 wireless sets in operational use.

Within five years the Royal Navy had doubled this figure and by 1905 it was totally dependent on wireless for its operations with over 80 ships fully equipped with long range telegraphy equipment. By use of the 'transatlantic' station at Poldhu Point, the wireless systems that grew out of the Boer War experience allowed the Admiralty to reliably communicate with any ship of the Royal Navy throughout the Atlantic Ocean, North Sea and Mediterranean.

As the first Royal Navy contract started in 1900, naval dockyards often had several Marconi men fitting and operating ships' sets with others undertaking experimental work on a small scale at the coast stations. Marconi also had his loyal band of dedicated experimental engineers and constructors. These groups of young men all worked with energy and enthusiasm and with a small factory in full production at Hall Street in Chelmsford they formed a loosely knit and flexible organisation. Marconi engineer M. Knight remembered that the apprentices regularly used to compete with each other by climbing up the Hall Street Factory's large aerial mast.

New recruits to the staff learnt 'on the job' from more senior men but it was realised that a haphazard training of this kind was not very satisfactory. The men directly involved with experimental work were well taught, but those on shift duty or at isolated coast stations were not so fortunate. In those early days, however, there were more jobs than men to carry them out, so that this disparity in training was partly discounted by moving staff from one job to another every month or so. As the organisation continued to grow, men often remained for longer periods in remote stations without transfer.

In December 1901 at great risk to his Company and reputation, Marconi gambled everything on what he called 'the great adventure' or 'the great leap' (*il grande salto*).

On 12th December 1901 he successfully received the Morse letter S, sent from a huge transmitter station at Poldhu Point in Cornwall while seated in a windswept tower in Newfoundland. The story will be forever famous. 'An uncertain kite flying in a giant storm and thought passed between' wrote one commentator.

At that moment the world changed. Marconi had forced a wireless signal over 2,000 miles across the Atlantic Ocean, and for the next ten years Marconi and his company struggled to develop the expensive and technically challenging transatlantic service.

By June 1905 Marconi's equipment had been developed so that wireless signals were being received at Poldhu from the Glace Bay station across the Atlantic, both stations being in daylight on a wavelength of around 3660m. While making improvements at Glace Bay, Marconi had recently devised a horizontal inverted L aerial which was found to have marked directional properties. However, as it was impossible to extend the aerial system on the headland at Poldhu, he decided to build a new station on the west coast of Ireland which would incorporate every new device known to the Marconi engineering team and take over the transatlantic service.

On 25th July 1905 Marconi and two companions travelled to Cashel, County Galway to inspect the Irish coast for a likely location. Shortly afterwards, a site 20 miles from Cashel at Derrygimla, about three miles southwest of Clifden was chosen and preliminary building work commenced in October of the same year. During the summer and autumn of 1906, the huge station grew and when completed was described as a shanty town, the whole 300 acre site was surrounded by a high barbed wire fence apart from where natural obstacles acted as boundaries. The entrance from the railway had a wooden sentry box with guards on duty day and night.

At its height the new Marconi station employed a permanent staff of 150 (including 10 engineers and 25 operators) with a further casual workforce of around 140 people with 70 men working full time to dig the peat. The station operated 24 hours per day with three 8 hour shifts. All the staff were male apart from a few housekeepers.

In 1907 both Glace Bay and Clifden were also equipped with a new form of disc discharger which overcame many of the problems of the old spark transmitter at Poldhu. Initial tests by both day and night began on 15th October 1907 and some 10,000 words were faultlessly exchanged to the great relief of the Marconi engineers. The inaugural message on the 15th October was sent at 11.30 a.m. from Lord Avebury to *The New York Times*.

At 9 a.m. on Thursday 17th October the first official messages came in from Glace Bay. These 'Marconigrams' or 'Aerograms' were charged at 5d per word for ordinary messages and 2.5d for press reports; this was half the charge of the transatlantic cable system. Reception rate was reported to be 30 words a minute in Morse code. The term 'Marconigram' continued to be used worldwide long after Marconi's had relinquished any control over the service.

The unlimited service began in February 1908 and in the first five months over 68,000 words in the form of 'Marconigrams' were efficiently transmitted for *The New York Times*. So useful had wireless become to the paper that it boasted: 'The first and only newspaper to use the trans-atlantic wireless telegraph, by which it receives daily more than 2,000 words from Europe.' The dispatches were all marked, 'By Marconi Wireless Telegraph.' The transatlantic service was an early success; 100,000 words were carried within three months of its initial commercial launch. Customers were not entirely satisfied, however, since messages were subject to delay with the Marconi Company blaming congested landlines serving the coast stations at either end.

The Company responded by announcing plans for upgrading Poldhu and Cape Cod to relieve pressure on Clifden and Glace Bay and to run in parallel with these stations. The four stations, it estimated, would then be able to process 20 words a minute, 12 hours a day, generating a net annual revenue of £150,000.

Marconi's dream of wireless communication across the Atlantic had come of age, and at last it seemed as if he might be able to challenge the cable operators. But it had been a cripplingly expensive process for his young Company.

CHAPTER THREE

Prelude and Disaster

The year of 1908 was to prove very difficult for both Marconi and his Company. The 'International Convention on Wireless Communication at Sea' had just come into effect which essentially ended the Marconi Marine Company's near-monopoly on supplying wireless equipment for merchant vessels.

Ship owners could now buy equipment purely on its technical merit, from any supplier, yet too many of them still failed to do so. Also the transatlantic wireless message service using the huge new Marconi stations that had taken a fortune to develop was still not yet profitable.

Marconi's had been in existence for twelve years and despite Guglielmo Marconi's many successes and world-wide fame, his Company was at a very low ebb. Over £500,000, nearly the entire cash reserve had been sunk in experimental work and no share dividend had yet been paid on the ordinary or preference shares and none was in sight. During 1908 the price of the Company's ordinary shares had dropped to 6s 3d and holders and investors who had paid £3 or £4 were not in a mood to make any further cash injections.

In another ill-judged venture in 1905, the Company had attempted to move into new manufacturing premises, when Cuthbert Hall, by then the Company's new Managing Director decided that Hall Street was overloaded and a new larger factory should be located closer to London. The Dalston Street Works in Tyssen Street, North London, was opened by Marconi's in 1905 and occupied the Shannon Factory, a very impressive building now called Springfield House that had been built in 1902 for a cabinet manufacturer. Much larger than Hall Street, the property had a total floor space of 31,170 square feet, valued at £61,585 in 1908, and included 'electrical power installation, lighting, heating and ventilation plant as well as fire appliances.'

The Marconi Works Dalston

Dalston Works Interior

The four storied building had three wings and was equipped with machinery removed from the Hall Street Works in Chelmsford. It took over the role of Hall Street in the manufacture of coils for the Marconi spark transmitters and paper capacitors. It later began mass production of ignition coils for use in the expanding automobile industry, but it was to be a brief career as the Dalston Works soon began losing money and was eventually closed down in 1908. In March 1908 Marconi himself had to step in to replace Cuthbert Hall and under his direction the original factory at Hall Street which had been closed three years earlier was re-opened.

But like many scientists Marconi found routine executive responsibilities extremely irksome. Nonetheless he had no alternative but to pursue an active policy of retrenchment after taking over as Managing Director to save the Company from the financial crisis that threatened to engulf it. With his usual skill and genius he turned the fortunes of his Company around, and by the end of June 1909 Hall Street was busy once more with orders in hand worth £87,000 (today worth around £8 million). But Marconi, tiring of the financial side of the business continually sought a replacement Managing Director that would free him from the day to day operational decisions of the large Company, and allow him to return to his first love of pure research and development.

Godfrey Isaacs then appeared on the scene. Just turned forty, he was one of the younger sons of the large family of Joseph Isaacs, a fruit-broker and general merchant with a large and profitable business. Godfrey had no obvious qualifications for the post Marconi needed to fill since he had no knowledge of wireless telegraphy or the new industry. But his father's business, which he had entered as a young man, maintained links with exporters on the continent, and Godfrey, who had been educated at Hanover and Brussels University, spoke several languages and had extensive contacts within many of Europe's finance houses. He had tremendous energy and enterprise, and Marconi, to whom he was introduced by the inventor's brother-in-law, seems to have taken an immediate liking to him.

Godfrey C. Isaacs became joint Managing Director with Guglielmo Marconi in January 1910. In August, having proved himself to Marconi and the Company Directors, Isaacs he took over as sole Managing Director. Isaacs was, like Marconi, multi-lingual and international in his outlook and he immediately undertook a full review the Company's structure and operations.

In October, Isaacs formed 'The Marconi Press Agency Ltd', a subsidiary, which in 1911 produced the world's first wireless magazine, *The Marconigraph*, renamed *The Wireless World* in April 1913.

Dr. James C.H. Macbeth, a noted cryptographer and one of the early members of the Marconi Company wrote.

> '**Godfrey** Isaacs, Managing Director of the English Marconi Company, was the king-pin of the organization. He was an extravagant promoter

and he had an insatiable love for power; he was the salesman of wireless with the business strategy and enthusiasm necessary to promote such a radically new communication system. He revelled in acquiring telephone and electrical instrument companies to link them as subsidiaries of wireless. He was generally faced with litigation, and from that Marconi, who detested routine business and legal conflicts, suffered pangs. Yet he entrusted the business end of wireless and its promotion to Mr. Isaacs, who presided at the Company's meetings and usually at public functions. Speech-making and writing were sacrifices for Marconi; in either he was concise.'

Godfrey Isaacs applied himself energetically to the problem of capitalising the expansion of wireless telegraphy and started litigation all over the world to prevent further infringement of Marconi's patents. By 1912 Marconi had established companies in Russia, Spain, the Argentine, Canada and America, all using the same master patents and all associated with the British Company, which usually held the majority of the shares and placed Directors on their Boards.

Isaacs knew the Company had to grow to survive and he was determined to not only enforce the Company's patents, but also systematically 'take out' the Company's four main competitors. In England the British Radio and Telephone Company (BRTS) were marketing a wireless system developed by John Graeme Balsillie an Australian engineer and inventor. The British Insulated and Helsby Cable Company were marketing their own wireless system and a syndicate headed by Scottish electrical engineer Alexander Muirhead and Marconi's old rival Professor Oliver Lodge marketed their own system. But perhaps the biggest threat was the German Telefunken Company. This had been born out of an amalgamation of Germany's wireless Companies and the work of many of its early pioneers, even if much of this had been appropriated from Marconi during his early experiments.

In 1911 Marconi's started legal actions for patent infringement. The ruling in favour of the Marconi Company was issued in the case brought against the 'British Radiotelegraph and Telephone Company' for the infringement of patent No. 7777 of 1900. The courts decided in favour of the Marconi Company recognising the validity of his patent. Marconi also sued the Helsby Company in 1913 and won. In 1912 Isaacs also brought the exhausting 14 year legal battle with the German radio company Telefunken to an end when the German combine applied for a licence for use of Marconi's master patent. Then Isaacs finally resolved a legal muddle left by an old quarrel between Marconi and Professor Sir Oliver Lodge that dated back to the earliest days of wireless. Isaacs persuaded him to accept £18,000 for his patents and Company and join the Marconi Company as a scientific adviser. In doing so he also effectively closed down the competing Lodge-Muirhead Company.

Isaacs had efficiently cleared the field of all major competitors, but he could not have realised that 1912 was to be a both a year of tragedy and a major turning point in the fortunes of the Marconi Company.

As the New Year dawned, Isaacs was reasonably pleased with the progress of the Company. He had cleared the way of competitors and it looked highly likely that the British Government would soon sign a contract for the proposed Imperial Wireless system. In March 1910 Godfrey Isaacs had submitted to the Colonial Office an imaginative plan to link the British Empire by a network of wireless stations, and on 7th March 1912, Mr. Samuel, the Postmaster-General, wrote a letter accepting the Marconi tender. The future looked bright for Marconi's and 1912 was indeed to prove to be a turning point.

On 14th December 1911 the Norwegian explorer Roald Amundsen and his team became the first humans to reach the Geographic South Pole. The English explorer, Robert Falcon Scott had also returned to Antarctica with his second expedition, the Terra Nova Expedition, in a race against Amundsen to the Pole. Scott and four other men reached the South Pole on 17th January 1912, thirty-four days after Amundsen. On the return trip, Scott and his four companions all died of starvation and extreme cold and some of their bodies, together with journals, and photographs were discovered by a search party eight months later.

On 10th April 1912 the new British ocean liner RMS *Titanic*, the largest passenger ship of her time left Southampton on her maiden voyage for New York City. Equipped with the finest luxuries, the massive vessel was truly a spectacle, with an on-board swimming pool, a gymnasium (complete with an electric exercise horse), squash court, Turkish bath, veranda cafe and libraries on both the first and second class decks. Each room was decorated with beautiful French polished mahogany furniture. In addition, the *Cafe Parisien* offered first class guests a dining experience unlike anything ever seen on a cruise ship.

She was also equipped with steam-powered generators, an electrical subsystem which provided lighting to the entire ship and two Marconi wireless systems. She was the pride of the White Star Line and joined her sister ship, the RMS *Olympic* which had already enjoyed great success and acceptance by the travel industry. The *Titanic* was also considered to be 'virtually unsinkable'.

On 11th April *Titanic* arrived at Queenstown, Ireland (today known as Cobh) picking up her final complement of passengers before steaming westward for New York. On 14th April the *Titanic* struck an iceberg in the northern Atlantic Ocean at 11:40 p.m. At 2:20 a.m. on 15th April 1912, less than three hours later, the *Titanic* sank beneath the freezing water, taking with her the lives of more than 1,500 people.

The sinking of the *Titanic* and the huge loss of life shocked the world but the tragedy would have been much greater had not 711 people been rescued, plucked out of their life rafts in the middle of the freezing ocean by a ship summoned by wireless while she was over 58 miles away.

In 1912 communication by wireless at sea, especially between ships, was still in its infancy, even though Marconi had established a transatlantic wireless message service four years earlier.

The first five years of Marconi's career had been a constant battle to develop wireless communication. He had faced massive engineering challenges, strong competitors, customer indifference, lack of funding, political and industrial espionage and rejection of his ideas by the established Victorian scientific community. But through it all Marconi had persevered and made wireless a reality.

On Wednesday 28th February 1900, the German liner *Kaiser Wilhelm der Grosse,* carrying 1,500 passengers became the world's first commercial merchant ship to be fitted with Marconi wireless equipment. But by 1912 fewer than 400 ships had been equipped and the intent of most wireless installations at sea was only to profit from the transmission and receipt of messages. Thus, the *Titanic*, like other large ocean liners was equipped with Marconi wireless systems primarily for handling wireless message traffic for revenue. It was the responsibility of the wireless operators, all of whom were employed by Marconi's and who sailed with the ship and their equipment to transmit and receive messages known as 'Marconigrams'. These included personal messages, general greetings, stock exchange quotations, business communications and news services. The possible use of wireless for signalling distress and emergency was incidental.

The sentiment of the period was that ships not carrying passengers (unless they were part of the Royal Navy) simply did not need to be equipped with wireless equipment. Those that were usually operated a normal daily schedule in line with their passengers' demands; hence 24 hour cover or emergency watches were unknown. When the wireless operator went to bed the equipment was simply turned off.

Although Marconi had been convinced since the earliest days of his experiments that wireless was essential for safety at sea, he faced a continual battle to get wireless systems accepted. The absence of any coherent regulations governing both safety of life at sea and wireless operation undoubtedly contributed to the *Titanic* disaster and the huge loss of life. After the *Titanic* struck the iceberg, the importance of the wireless room and the ability to communicate between ships was recognised as being essential.

On the evening of 14th April 1912 the night was clear and starlit. The sea was calm. This was a rare event in the North Atlantic. These conditions were a sailor's dream and *Titanic*'s Captain E.J. Smith apparently saw the opportunity to go full speed ahead in an attempt to break the ocean crossing speed record (the Blue Riband) held by the rival Cunard liner the RMS *Mauretania* since 1909. Smith was an extremely experienced sea Captain and Master for all the White Star Line's maiden voyages. After *Titanic's* first voyage he was due to retire and it has been speculated that in pursuit of the record on his last command he may well have thrown caution to the wind. This may account why the *Titanic* apparently failed to heed ice warnings received by wireless.

At 11:40 p.m., a lookout repeatedly shouted the warning of 'iceberg dead ahead!!'

The bridge attempted evasive action and full reverse propellers was attempted, but it was too late. The inertia of the *Titanic* was too great and she continued to plunge ahead on a path that ripped open the sealable bulkheads on the starboard side, a rupture estimated to be about 300 feet in length. The seemingly invincible *Titanic*, having lost its sealable airtight bulkheads was no longer able to remain buoyant at its bow. Within two and a half hours of striking the iceberg, the bow submerged. At 2:10 a.m., the massive stern rose into the air and the ship plummeted to a watery grave, carrying with it 1,513 helpless and terrified passengers.

Earlier, aboard the SS *Californian*, the Marconi wireless operator Cyril Evans had turned on his wireless equipment to clear his routine traffic. But being only ten miles from the *Titanic*, the operator on duty on the *Titanic* strongly advised Evans to 'shut up', as he was interfering with their commercial wireless traffic to the Cape Race wireless station in Newfoundland. Evans complied. Being the lone wireless operator on the *Californian* and having worked a long day, Evans switched the equipment off and retired for the night. The *Californian*, within sight of the *Titanic*, had found itself in the same ice field earlier in the evening at 11:00 p.m. Wisely, Captain Arthur Rostron of the *Californian*, ordered his ship to a complete halt, intending to carefully wend his way out at daybreak.

The *Titanic* struck the iceberg at 11:40 p.m., less than a minute after the first sighting of the iceberg by the lookout. But the first 'CQD', (General Call Distress) was not initiated until 12:15 a.m., thirty-five minutes later. The *Californian*'s First Officer observed white flares shot into the sky from the *Titanic*. Unfortunately, he assumed the flares to be part of a celebration aboard the *Titanic* as was the custom during heavy partying. He also considered the possibility that he was observing shooting stars. In 1912 the arbitrary discharge of white or coloured flares was acceptable and commonplace as there were no regulations governing the deployment of flares as a signalling or emergency tool. The uncertainty of the First Officer nevertheless prompted him to use a Morse code signalling lamp aimed at the *Titanic*, but he received no light signal response. Neither the First Officer nor the Captain of the *Californian* attempted to wake Evans and direct him to the transmitting key of the wireless to send a message of inquiry to the *Titanic*. With this one failure aboard the *Californian*, within sight of the *Titanic*, the fate of the 1,513 lives was sealed.

Fifty-eight miles to the southeast of the *Titanic* was the SS *Carpathia*. The ships wireless operator Thomas Cottam was also preparing to retire when by chance he initiated contact with the *Titanic* to advise its operator that the Marconi land station at Cape Cod was attempting to contact him. The response from the *Titanic* was prompt, with an urgent message distress message and requesting immediate aid. The *Carpathia* turned its course 140 degrees and headed for the *Titanic* at full speed. All together, eight ships over a wide area heard the *Titanic's* 'CQD' distress call and were racing to the scene, including the *Frankfurt* from over 140 miles away.

The *Carpathia* was the first to arrive at the scene of the disaster at 4:15 a.m. There was no *Titanic*, only an empty sea dotted with lifeboats, adrift without lights and the shivering passengers in them huddled together to protect against the freezing air. By 8:30 a.m. all survivors had been picked up, The *Carpathia* had recovered 14 lifeboats and 712 survivors, although one of them died later en route to New York.

By then, the world was waking up to the news that the unsinkable ship had been lost. Families with loved ones aboard the vessel were desperate for a list of survivors, but it wouldn't be compiled until a week after the accident. One can only imagine the despair of the *Californian* crew in the morning when they were told by wireless of the *Titanic's* loss. They were only a few miles away but they were the last to know when the wireless operator returned to his station in the morning to begin routine traffic operations.

The man who gave the world wireless had also nearly been lost in the *Titanic* disaster. Marconi and his family were invited by the White Star line's Chairman Bruce Ismay to be his guests on the maiden voyage of the *Titanic*. Fortunately for Marconi he had to hurry to America on business, so he went earlier on the RMS *Lusitania*. Travelling with the Managing Director Godfrey C. Isaacs and his son Marcel, Marconi had arrived in New York on 15th March. They had proceeded to the United States in connection with the affairs of the American Marconi Company, which was controlled by the English company. In addition, as part of Isaacs world-wide purge Marconi's had brought a legal action against the United Wireless Company for patent infringement that was being heard in the Federal Court on 25th March. The Company was in liquidation and with some of its directors in prison the English Marconi Company purchased the assets of this company and resold them to the American Marconi for 1,488,800 fully-paid shares of $5 (£1) each in the latter Company.

Marconi's wife Beatrice had however retained her booking on the *Titanic*, but on the eve of the voyage, she too cancelled due to their son Giulio's sudden illness with a high fever. She cabled Marconi that she had to postpone her trip and the *Titanic* sailed on her fateful voyage on 10th April without the Marconi family aboard. Marconi still held a ticket for the *Titanic's* return voyage to England, planned for 20th April 1912.

Marconi arrived in New York just in time to hear that a wireless message had been received at the Cape Race wireless station in Newfoundland which might indicate a disaster at sea. The *New York Times* promptly sent a wireless message to the *Titanic's* Captain, Edward J. Smith. They got no reply.

A period of total confusion ensued. The full horror and tragedy of the disaster was only fully comprehended when the *Carpathia* sailed into New York's Harbour through the rain on Thursday night. As soon as her gangplank went down, Guglielmo Marconi stepped out of the immense and silent crowd at Pier 54. With police clearing the way, he was one of the first people along with Mr Speers of *The New York Times* to go aboard to interview the wireless men whose work had saved so many. The *Carpathia's* wireless operator Thomas Cottam and *Titanic's* second wireless officer, Harold Bride were onboard. The *Titanic's* first wireless officer, John George Phillips, had been lost in the disaster.

Suddenly wireless was the hottest news topic in the world and the New York Electrical Society invited Marconi as a guest lecturer on 17th April 1912, just days after the disaster. The loss of the *Titanic* was the main headline in all newspapers. Marconi, praised as the saviour of 711 lives was in the forefront of the news as never before. It seemed that all New York wanted to see him before he returned to Europe and the Engineering Societies auditorium was jammed to capacity. When the inventor appeared at the side of the platform, the crowd in the balcony saw him first, and the cheering began. It spread to the main floor and was continuous as Marconi bowed many times.

On 18th June 1912, Marconi started to give evidence to the Court of Inquiry into the loss of the *Titanic* regarding the marine telegraphy's functions and the procedures for emergencies at sea. The principal finding was inescapable. Without wireless on board the *Titanic,* all 2,224 passengers and crew would have perished. The popular press of the day hailed Marconi as a hero. A tribute to Marconi printed in *Punch* Magazine after the *Titanic* disaster saw the cartoon character Mr Punch doffing his hat to the Italian pioneer, with the following caption:

'Many Hearts Bless You Today Sir, The Worlds Debt To You Grows Fast.'

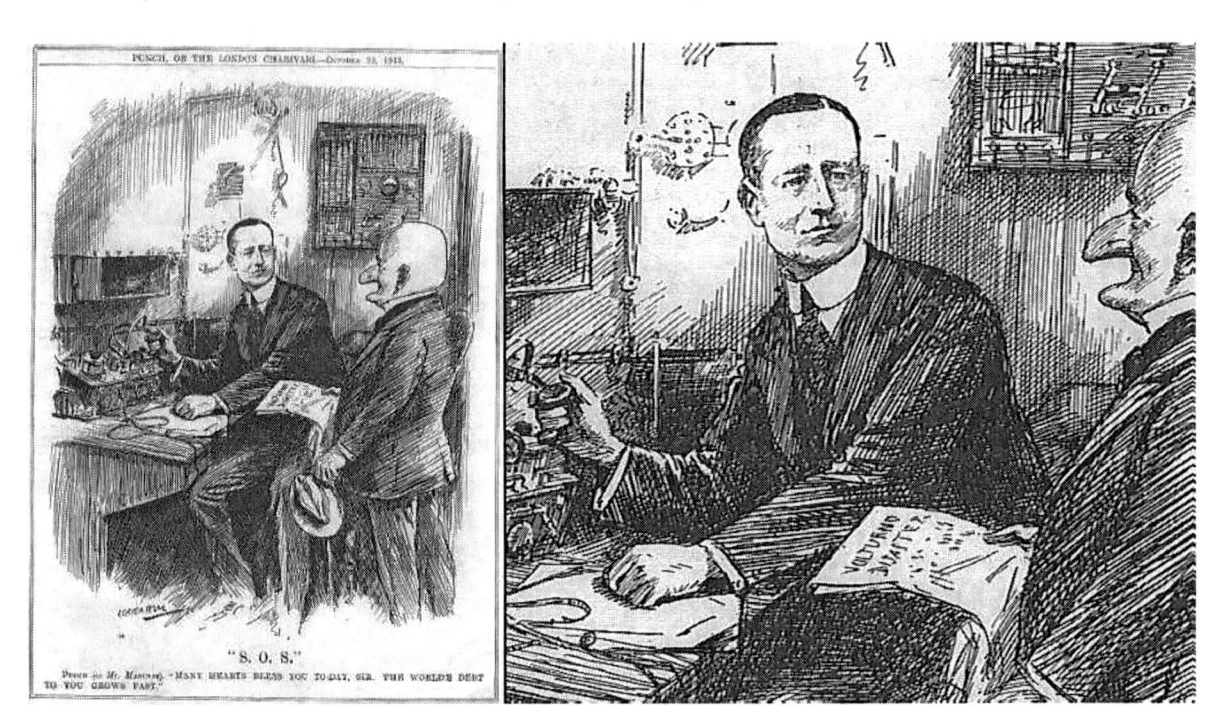

Mr Punch honours Mr Marconi

As Lord Samuel, the British Postmaster General at the time, stated: 'Those who have been saved have been saved through one man, Mr Marconi and his wonderful invention.'

A few days later the survivors of the *Titanic* presented Marconi with a solid gold medal, in gratitude for Marconi's wireless installation on board the *Titanic* credited for saving their lives. The survivor's cries of *'Ti dobbiamo la vita!'* [*We owe you life* !], remained with Marconi for the rest of his life. Four weeks after the *Titanic* disaster honours and editorial tributes were still being heaped upon Marconi, including on 21st May 1912 the award of The Grand Cross of the Order of Alfonso XII from Spain.

It is possible that the *Titanic* disaster could have been averted had the lessons been learnt from an earlier accident at sea. On 23rd January 1909 at 5:30 a.m. the luxury liner S.S. *Republic* had left New York with 461 passengers bound for the Mediterranean, was rammed amidships by the S.S. *Florida* in a dense fog about 26 miles southwest of the Nantucket Lightship. Fortunately, the S.S. *Republic* was equipped with a Marconi wireless system, while the S.S. *Florida* was not. Following the impact, Jack Binns the Marconi wireless operator on board the *Republic*, immediately started to send 'CQD' distress signals and many ships rushed to its aid. For two days in freezing conditions Jack Binns sent out a total of two hundred messages to help guide rescuing ships to his stricken vessel's position.

All 461 passengers, except for five crew members survived the disaster and were transferred to the S.S. *Florida*. The SS *Republic* was abandoned and the SS *Florida* herself in danger of sinking transferred all passengers to the S.S. *Baltic* for the return trip to New York. The combined number of passengers of the two ships *Republic* and the *Florida* totalled 1,650. Although hailed as a great triumph for wireless communication, the small loss of life meant that there was no public outcry following the near disaster. Thanks to wireless all the passengers and crew who had not been killed by the initial impact were rescued. Jack Binns received a special medal for his services and Marconi himself presented him with a gold watch.

But after the *Titanic* went down, the ensuing inquiries both in Great Britain and the United States led to far reaching legislation including a requirement that every ship in the world over 1,600 tons had to carry wireless.

The most significant result of the disaster investigations was the call for an International Radio-Telegraphic Convention, to convene in London, on 5th July 1912, for the purpose of establishing regulations and procedures governing wireless services aboard ships and shore stations. Attended by sixty-five countries, regulations and procedures were enacted, some of which are still in effect today. Among these is the use of 'SOS' as the universal call of distress as it was determined to be the simplest form of signalling to replace 'CQD'. The abbreviated 'Q' code signals that are still in use today were also an outcome of the meeting.

The 'Safety of Life at Sea' Conference was held in London on 12th November 1913 and was attended by sixty-five countries. This conference was the turning point for wireless communication at sea. Sweeping regulations were put into effect governing all ships at sea, whether motor propelled or under sail. Ocean going vessels carrying passengers to foreign ports were mandated to be fitted with a wireless communication system. Further, the ship's wireless room and shore stations were to be manned twenty-four hours a day. Now the wireless room became the focal point on board with all vessels having to abide by all new rules, regulations and laws, establishing safety of the passengers and ship as the first priority. The value of wireless on board was now self-evident.

The 'Safety of Life at Sea' Conference concluded on 20th January 1914. It was determined that all countries having ocean going vessels carrying passengers were culpable for inadequate safety regulations on its vessels. The Conference emphasised the necessity for united action to revise the old laws and adapt them to new conditions.

On 9th October 1913, Wireless was again instrumental in the saving of 650 lives from the SS *Volturno*, an ocean liner that burned and sank in the North Atlantic. She was a Royal Line ship under charter to the Uranium Line at the time of her fire. At about 06:00am the *Volturno*, carrying mostly immigrants bound for New York, caught fire in the middle of a gale in the North Atlantic. The crew attempted to fight the fire for about two hours, but, realizing the severity of the fire and the limited options for dousing it in the high seas, Captain Francis Inch had his wireless operator send out SOS signals. Eleven ships heeded the calls and headed to *Volturno*'s reported position, arriving throughout the day and into the next. Wireless had done its job, but tragically several of *Volturno*'s lifeboats with women and children aboard were launched in heavy seas, all the boats either capsized or were smashed by the hull of the heaving ship, leaving no one alive.

This near disaster after the tragic sinking of the *Titanic* in 1912 re-focused public attention on the importance of ships being equipped with wireless and the necessity for maintaining a 24 hour communications watch. Many countries now required that all ships over 500 tons be equipped with wireless. This caused demand for marine equipment to soar. Within a decade, over 500 shore stations had been built, establishing wireless (now almost universally known as radio) communication worldwide.

The Marconi Company share price that had been languishing at an un-saleable ten shillings or less, rocketed almost overnight to £10.00 and in 1913 the Marconi Company was finally able to declare its first dividend to its shareholders.

It was the tragedy of the *Titanic* that gave birth to the modern wireless age and spurred a growth in manufacture and development that probably would not have occurred otherwise. What better advertisement for the 'wireless' could there be than a world-wide news media who printed the disaster story day after day for months?

The mystique and the magic behind the word 'wireless' also gave birth to a new generation of aspiring operators and engineers, together with the need for accelerated manufacture of wireless equipment to fulfil the demands of ship and shore installations. The tragedy of the *Titanic*, occurring when it did during a period of slow growth in a new industry was responsible for the kick start of the wireless, radio and electronics industry that today still provides the greatest number of jobs in the history of civilisation.

The Marconi Company recognised the need for operator training and established Marconi Wireless training schools throughout the world, including the major cities of the United States. The new regulations requiring wireless on board all ocean going vessels made it necessary for the Marconi Company to significantly step up equipment production to meet this need.

To do this the Company urgently needed new manufacturing facilities on a scale never before imagined in the still embryonic industry.

All this was the legacy of the **Titanic** disaster.

CHAPTER FOUR

NEW STREET

The World's First *Purpose* Built Wireless Factory

Even before the *Titanic* set sail, plans for expansion within Marconi's were already underway. It was clear to Isaacs that the original works at Hall Street was becoming too small as the demand for new wireless telegraphy equipment had already increased tenfold. A new wireless station had been constructed in Broomfield on the outskirts of Chelmsford in August 1903. This was first used for tests with the new Poldhu station's 'T' aerial operating on a wavelength of 2,000 meters. The Broomfield site provided with Marconi a research station to study wireless transmission with powers somewhere between the huge Poldhu station and the limited power of marine equipment. The station was headed by senior engineer Captain H.J. Round, but there was insufficient space in Broomfield for any form of major expansion.

By the end of 1910, Marconi was able to prove beyond doubt that wireless could provide an alternative to long-distance cables between fixed points. With 600 patents and equipment installed in a total of more than 500 wireless stations, its reach was now almost worldwide.

The Hall Street Works were becoming seriously cramped. Double shifts were being worked in the factory to make wireless equipment for export to the four corners of the globe. Customers included the Amazon basin, Thailand, South Africa, India and even to both sides in the Balkan War of 1912. Even the trans-Atlantic cable companies, Marconi's greatest competitors, were customers.

Marconi wireless equipment was also the corner stone of the growing number of shore based wireless stations and his equipment was carried aboard all the great Atlantic liners including the *Lusitania, Mauretania, Baltic, Olympic* and the ill-fated *Titanic*.

Having learnt from their mistakes with the ill-advised move to the Dalston factory in 1905, in January 1912 Marconi's Managing Director Godfrey C. Isaacs proposed building the world's first-ever *purpose-designed* and *purpose-built* wireless factory on the local cricket ground in Chelmsford that was owned by the Church Commissioners. The proposed new works would cover the whole of the site. To the north two new roads would be constructed leading from New Street parallel to Rectory Lane (Marconi Road and Bishop Road) where cottages would be built for the Company employees.

Marconi's commissioned the architects William Dunn and Robert Watson in London to draw up plans for the first factory to be specifically designed for the construction of Marconi's wireless equipment. Initially the architectural practice operated from 35 Lincoln's Inn Fields and by temperament, Dunn was more a mathematician and structural engineer than an architect and he had a particular flair for design in concrete, while the decorative aspects of the partnership's work fell to Watson.

Godfrey Isaacs, wanted the new factory to be finished and working by mid-June 1912, an almost impossible target, but he wished to show off his smart new wireless factory on 22nd June to his leading competitors, Government officials and other experts who would then be in London for the Wireless Conference.

Godfrey Isaacs's plans were in fact not just for a new factory; he wanted the new complex to be a complete self-contained village within a town and an agreement was made to purchase the local Cricket Ground. Chelmsford Cricket Club could trace its history back to 1811 and in 1879 the Club had moved to the land adjacent to New Street which became their home for the next 32 years. But the club now became a victim of Chelmsford's rapid industrial growth.

As the area was Glebe land, adjacent to the Rectory on Rectory Lane and belonged to the Church, it had to be purchased from the Bishop of St Albans, the Rector of St Marys and the Ecclesiastical Commissioners. Once the contract was signed the cricket ground was pegged out on 10th February 1912, and the bricklaying using 500 men, started on 26th February. Despite a short building strike, just seventeen weeks later, the changeover from Hall Street to the new 70,000-square-foot (6,500 m²) factory complex was accomplished in just one weekend, leaving the old Silk Mill once more empty and abandoned, although the wireless station remained operational. Amazingly all sections of the new works were fully functioning in time for the International Radio-Telegraphic conference.

The site was ideal for Marconi's plans. Located due north from Hall Street the ex-Chelmsford Cricket ground had an area of around 10 acres. The large railway coal yard opposite gave easy access to the main London Great Eastern Railway. Rail trucks could bring coke to feed the Company's power station, and the heavy components could be loaded directly onto trucks for nationwide delivery or to the docks for shipment overseas.

No expense was spared in the works design; arches were prominent, both internally on office doors and on the south side of the building facing the courtyard. The new works were designed from the outset for mass production and equipped with the latest and best tools, apparatus, test rooms and laboratories. The factory frontage ran alongside New Street with a clock tower above the main entrance. This building housed the Company offices, drawing office, and showrooms. The workers entrances, (men and women had separate doors) were to the left while a separate building contained the separate male and female dining areas and clubrooms. Typically men worked on machine tools in the machine shop or in carpenter's shops, whilst the women worked producing induction coils for the spark transmitters.

Behind the office block was the main test area, above which six glass cupolas provided light. Behind this was the factory where overhead line shafts were driven by DC generators, supplied in turn by state of the art turbines in the power house. In the main yard a single 200 foot wireless mast was erected to support an antenna for transmitter testing.

During construction the building required 2.5million bricks, 400 tons of steel and 9,000 truckloads of earth were removed. Sewers were diverted and a 400 foot well was sunk. The completed factory frontage was 200 feet long and 40 feet wide and the whole site was fitted with low pressure hot water radiators throughout. Railway lines were extended across New Street from the main shunting yard and the national network beyond and new sidings were constructed on one face of the works with two tracks, one in and one out with a drive way running between. The packing department had two loading bays with turntables and two electric capstans to winch the carriages in and out of the works so they could be loaded and unloaded under cover. The factory building arches were built to accommodate the exact size of the railways carriages so they could be manoeuvred in and out for loading. At the entrance to the rail sidings there were two weighbridges checking the weight of raw materials delivered. The factory was designed to be self-sufficient where ever possible and to manufacture all its components and sub-assemblies in house. Hence wood and iron, steel and packing materials would continually arrive and completed equipment departed via the railway.

The main New Street Works building behind the New Street frontage was 466 feet by 150 wide with its roof constructed to a 30 foot saw tooth spacing design with 65 wooden shafts to give ventilation. It was capped with green slates on felt and match boards. The whole glazing system used was Rendle's 'Invincible' glazing, normally used in train stations and large public buildings. The system used steel T bars with specially shaped copper water and condensation channels, all formed in the one piece and resting on top of the T steel; the glass rests on the zinc channel, and a copper capping is fixed over the edges of the glass and secured with bolts and nuts.

The factory floors were tongue and groove Ash wood blocks laid over the concrete floor, while the walls were Fletton bricks. The Fletton Brick Company eventually became the London Brick Company and the dominance of London Brick in the market during this period gave rise to some of the country's most well-known landmarks, all built using the ubiquitous Fletton.

The power test area had granolithic floors due to the heavy machinery involved along with a five ton, three motor overhead travelling crane on a runway that extended over the loading dock in the packing department. Fire risk precautions were taken very seriously in the new factory. Fireproof doors separated each department and the entire building had been installed with automatic glass disc

Grennell sprinklers. Although commonplace today, sprinklers were a new safety innovation in 1912 but considered essential in a factory were spark transmitters, high power electricity, generators, wood working and packing materials all existed practically side by side.

The builders selected for the project were Cubitt & Co Ltd (actually Holland, Hannen & Cubitt's Ltd) a major building firm based in London who were responsible for many of the great buildings of London. The Company had been formed from the fusion of two well-established building houses that had competed throughout the later decades of the nineteenth century but came together in 1883 when Holland & Hannen acquired Cubitts, a business founded by Thomas Cubitt some 70 years before. The combined business went on to construct many important buildings and structures including the Prudential Assurance building in High Holborn (1906), the Cunard building in Liverpool (1917), the Cenotaph in London (1920), London County Hall (1922) and South Africa House in London, completed in 1933.

The New Street project was so successful that the Company awarded the same building company construction of its new long-wave transatlantic transmitting station near Waunfawr, on the slopes of *Cefn Du*, three miles east of Caernarfon. The first land was purchased in December 1912 and work started on the 21st February 1913, the main building measuring 100 feet by 50 feet and elevated 830 feet above sea level. In 1915 the Admiralty required a powerful spark transmitter to be built at Moody Brook to the west of the Stanley settlement in the Falkland Islands. The contractors to the Admiralty were Marconi's who again sub-contracted construction to Cubitts who employed 20 labourers and duly despatched them to the Falklands.

The 250 International Radiotelegraphic Conference delegates made their inspection tour of the brand new Marconi New Street Works on 22nd June 1912.

A special train brought delegates to Chelmsford from London for the grand opening of the new factory. The Mayor of Chelmsford, Alderman T.J.D. Cramphorn, J.P, accompanied by his sister, the Mayoress, stood in the new entrance hall to welcome Marconi, his team and the world's leading wireless experts. It was a great honour to have all the delegates gather in his Borough, including representatives from as far away as Egypt, Japan, Turkey, Morocco and Siam.

The works tour included a full working demonstration of the new Marconi Wireless Telegraphy system. From the main hall over the polished wood floors the visitors were led to the left past the offices, where they were taken outside to the far side of the factory and the factory tour began at the carpenters' shop where woods such as mahogany and teak were being crafted into mountings for the wireless components. The visitors were then taken into the machine shop, at the time one of the finest and largest in Essex, 187ft. long by 90ft. wide. Here large D.C. motors powered two overhead line shafts driving the individual machines.

From here the visitors continued through raw stores, where tons of ebonite and brass were stored. The next stop was at a point below the water tower that housed a large 8,000 gallon tank connected to the Grinnell fire sprinkler system. The tank was filled from the 400 foot borehole dug into the chalk that was lined with 6 inch steel tubes to a depth of 333 feet. On their whirlwind tour of the new works the visitors were next taken outside to the railway sidings where railway trucks were delivering the raw materials and despatching completed wireless telegraphy equipment. Opposite the railway siding could be seen the powerhouse containing a number of steam turbines including a 27,800 r.p.m., 45hp turbine. From here the delegates were taken through the packing department where the completed wireless equipment was packed for shipping; Finished Stores for items awaiting packing; and finished parts stores, where sub-assemblies were housed; to the condenser and winding shop, the winding shop being staffed entirely by women.

The tour continued on to the mounting shop where wireless telegraphy sets could be seen on an assembly line, being mounted onto army carts. One of the problems during the Boer War in South Africa had been the cutting and tapping of the telegraph wires by enemy raiding parties. The new Marconi Wireless Telegraphy system of course overcame the problem and many of the carts shown to the delegates were already in use in Tripoli by the Italian Army. From the mounting shop the delegates were ushered to the test room, a long shop, situated behind the main façade alongside New Street, daylight entered the room from the six cupola skylights along its length. On occasion the discharge from a spark gap transmitter on test sounded like a rifle shot and made each group on the delegate's tour jump with much nervous laughter.

New Street Breaking Ground Ceremony

Foundations Underway

Part Built

The New Street build team

Scaffolding to main building

The Completed New Street Works

Visit by International Radiotelegraphic Conference delegates

Marconi personally hosted the New Street Works
visit of the International Radiotelegraphic Conference

New Street, c1915

New Street Power House, c1920

New Street Main Yard, c1915

New Street, c1920

The tour still continued via the power and oil test departments back to the main building and the showroom where many examples of the latest Marconi wireless telegraphy equipment were on show. During the tour, visitors could see a replica of a ships wireless cabin; military equipment including cavalry wireless apparatus, carried by horses, various valve receivers, a 3kW set with quick-change tuning of the primary circuits, a 5kW battleship set, a 1½-kilowatt ship set, a new ½kW cargo set, and also a wireless equipped car, where an operator communicated with Hendon Airport.

The Company had spared no expense to make the tour the showcase for all that the new technology and the new factory could offer. Marconi engineers had installed the new Bellini-Tosi direction finding system, by which the direction of a transmitting station relative to the stations receiving the message could be ascertained. The experimental system had only that year first been installed for trials aboard the RMS *Mauretania*. The highlight of the tour was a demonstration of transmission and reception between the new works and the Wireless Telegraph site at Poldhu in Cornwall. For this a 15 kW ship set was connected to the main aerial suspended from a 200 foot high temporary tubular mast outside. Transmission wavelengths ranged from 600 to 2,800 metres, the spark set emitting an almost 'musical' hissing note of 400 Hz.

After touring the works the visitors were taken past the temporary mast and on to the old cricket pavilion where a Morse sender and Morse inker tape machine had been installed. The following message was sent to the Poldhu Point station: 'The President and delegates of the International Radio-Telegraphic Conference present to the staff of the station their very cordial greetings.' Poldhu replied: 'To the President of the International Radio Telegraphic Conference, - The engineers and staff on the Poldhu station have the honour to present their respectful homage to all the delegates.'

The inspection ended in a large marquee, where the caterers Messrs. Hicks, Son and Co. provided an elegant tea, with ices, strawberries and cream. Marconi himself oversaw the whole proceedings, dressed in a dapper blue flannel suit with a fine white-stripe, and a straw hat.

In the evening Marconi and his Directors entertained four hundred guests to a magnificent banquet at the Savoy Hotel in London, a stone's throw from the Company's impressive new London headquarters, which had recently, and rapidly, been converted from an unoccupied apartment block in the Strand and renamed Marconi House. At the evening function the Mayoress of Chelmsford wore black satin, trimmed with pink chiffon. She and the rest of the guests were received by Marconi and Godfrey Isaacs and the party then proceeded to the impressive dining-room of white and silver, adorned with flowers and illuminated with concealed lights. From the gallery an orchestra played music from various nations.

After the meal the Marconi Company presented gifts to all the delegates, each lady receiving a silver perfume scent bottle, while each gentleman was presented with a silver cigar lighter, this firing a spark on to a wick charged with petrol. The lighter was in the shape of Marconi's famous disc discharger system that he had patented in 1907. The after dinner speeches were in French, but then Marconi spoke in English, stating that for the first time since the invention of wireless telegraphy, representatives of every country in the world had assembled first in Chelmsford and then in London to help to form International regulations governing the application of this discovery. Marconi said that this was for him, personally, a great honour and he had been pleased to receive them at the Chelmsford Works. He raised his glass to those who had assisted him thanking all present, and proposed a toast to the health of the delegates.

The following weekend, the delegates visited Poldhu as part of a weekend programme. The tour concluding with a garden party at 'Eaglehurst', Marconi's private residence on the Solent. The conference was a huge success and the New Street factory soon became a busy and vibrant manufacturing plant. The *Titanic* had sunk in April while the factory was being finished. The new International regulations that followed, bringing some good out of the disaster, meant that large volumes of new orders and work poured into the Marconi factory.

Under this new load even the huge New Street site started to get congested and Marconi's started to use army huts to house various new developments. These could be seen around the factory building, in the courtyard and to house the transmitter at the base of the antenna feed. One 'smelly' problem was quickly addressed when soon after the factory opened sprinklers were added to the work's pond to prevent stagnation.

Marconi New Street Works from the Air, c1920

The coal yard can be seen in the foreground on the right

Key

A	Club Room	N	Packing Case Shed
B	Men's Dining Room	O	Hospital
C	Women's Dining Room	P	Toilets
D	Kitchen	Q	Winding Shop
E	Office	R	Cooling Pond
F	Luncheon Room	S	Well House (Deep Well and Pump)
G	Reception	T	Ladies Clubrooms
H	Drawing Office	U	Test Huts
I	Showroom	V	Tool Room
J	Cottage	W	Machine Shop
K	Army Huts	X	Water Tower
L	Mounting Shop	Y	Packing Shop
M	Instrument Test	Z	Two 30 h.p. Austin Generators

a	105 h.p. Crompton motor & 3 phase generator	f	Saw Mill
b	Timber Store Shed	g	Timber store
c	Power House	h	Valve Testing
d	Marconi Houses	i	Petrol Pump
e	Transmitter Hut	j/k	450ft Aerial Masts

Machine Shop

Mounting Shop

Above and Below - The Winding Shop

Paint Shop

Raw Stores

Instrument Test

Carpenters Shop

Accounts

Drawing Office

Case Construction

Packing Shop

Examiners Department

Assembly Shop

Finishing Store

Finished Parts Stores

Surgery

Blacksmiths Shop

Above and Below - The Power Test Department

Power Test and New Street, c1915

Marconi Works, Chelmsford

Above - 15kW Transmiter in the Show Room, Below - The Show Room

Aerial feed to main aerial, c 1933

The Essex Chronicle in 1912 reported that the temporary 200 ft. mast at New Street was to be replaced by two '*500* ft' masts 'as soon as the castings are completed'. In October 1912 it reported that the masts were 'rapidly rising' and they were completed by May 1913. The July 1935 report on their being dismantled also stated that: 'They were erected in 1913'.

The huge new 450 foot high masts, set 750 feet apart at the base were known locally as 'the drainpipes'. Each mast was three feet in diameter and made from four sections of half inch pressed steel plate with vertical flanges bolted together. A single flange at each end joined each section together. Five sets of insulated stays with a radius of 220 feet connected each mast with four steel anchors set into 100 ton concrete blocks while at the base of each mast a huge concrete block of 120 tons supported the load. These huge steel tubes dominated the Company's New Street Works and quickly became unforgettable landmarks, visible across the town and whole area. There was a story that Frederick Post, the mast erector Foreman had climbed the mast using just the bolt heads to enable a replacement for a broken timber top section to be hauled up during WW1.

The Marconi Poles

'I was born in 1909 in Primrose Hill and I can remember my mother standing me on our kitchen table to watch the 'Marconi Poles' being erected behind the New Street Works (when I was about 3 years old)... We saw the steeplejack building up the quarter sections and going up and down on a cradle as each section was bolted in place.... Some years later there was a heavy snow storm in the spring term and the weight of the snow as it built up and froze on the twin aerials snapped one of the masts on the top of one pole, again a man had to climb up on the bolts (a few feet at a time) to repair the damage. Either in the late 20s or the early 30s these poles were dismantled and I can vividly remember watching from my office window in Waterhouse Lane the workman cutting off piece by piece with presumably an acetylene torch.

Gerald Hockley
[the poles were dismantled in 1935]

'**The** ones at Chelmsford were about 450 feet and I have actually got a bolt from one of them... Oh it's a hefty great thing... makes a lovely doorstop.... they used to climb up these masts in what looked like a sort of basket thing to gradually dismantle them and it was right at the top of the yard getting up towards Broomfield Road end.... we just went up and said could we have one and they just gave us one. I've often wondered how I got it home because I only had a bicycle basket.'

Doris Collision

'I was born in 1926 and as a very young child lived at the top of Marconi Road opposite to one of the 450 ft masts.

My mother recalls a lightning strike which brought down the aerial and saw a man climb up on the bolt heads to fix up a Bosons chair. A few years later I moved into Bishop Road and I can vividly recall watching the other mast being cut up. I was intrigued by the blue acetylene flame as the work proceeded.

It was during the mid-1930s that I would be sent down into The Works to have our wireless accumulator charged up and bring one back in exchange. I remember because the acid spilt and burnt my clothing!

At the bottom of Marconi Road were two semi-detached houses. One for the Work's Electrical Engineer and the other for the Works Mechanical Engineer. The latter became close friends with my parents. His name was George Barlow who I called 'Uncle Barr'! It is my belief that he was in charge at the commissioning of the diesel generator sets in the power house and subsequently 'stayed on'.

Brian Thwaite

Original 1912 New Street Mast, 250ft high

New Street Masts Construction gang

Marconi's New Strreet Masts

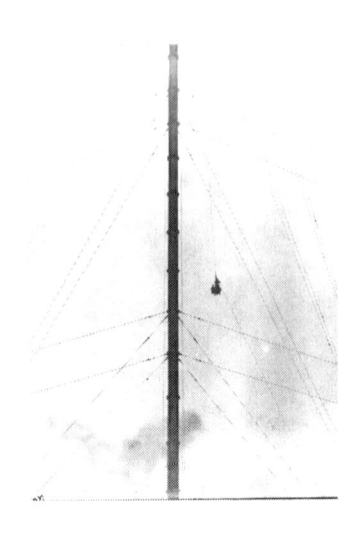

New Street Mast Demolition, 1935

Mast section - Cut up on the ground for scrap, 1935

The opening of the Marconi New Street Works in June 1912 had been a huge success, but in July 1912 Guglielmo's luck changed. In the wake of negotiations with the government to implement the Company's ambitious plan for an Imperial Wireless Scheme, the city was swept with rumours that ministers, privy to the information, had been speculating in Marconi shares. Subsequently the rumours were found to have no substance, but what became known as the *Marconi Scandal* left its scars on the Italian inventor, both emotional and physical. Marconi, somewhat disillusioned with Britain and the British Press left for his native Italy. Over the next 25 years he was to return on only a few occasions that suited his research and international lifestyle. He maintained no official office at the New Street Works and there are few references to him even visiting the site.

By the end of 1912 Marconi had crossed the Atlantic more than fifty times without accident of any sort and even missed the ill fated maiden voyage of the *Titanic*. His invention had added truth to the expression, 'man is safer at sea than on the land.' But the dangers lurking on the terra firma overtook him on 25th September 1912, when the motor car in which he was travelling from Spezia to Genoa in Italy collided with a car carrying several Venetian ladies.

At 12.30 p.m. Marconi was at the wheel of his car making good speed in order to climb a high curving road through the mountains. A quarter of a mile from Casa de Vara outside of Spezia the two cars crashed head on and rescuers found Marconi stunned and clinging to the wheel. A naval ambulance from Spezia rushed to the scene and Marconi was taken to the Naval Hospital in Spezia. All the naval and military surgeons available quickly gathered at his bedside. His right eyeball, right temple and cheek were badly bruised. The eye was cut by a splinter of glass piercing the eyeball. On 17th October 1912 Marconi lost his right eye. Italy was shocked at the news and messages of sympathy came from all parts of the world. For Marconi, the immediate future involved a long period of rest and recuperation.

Marconi was back in England at the outbreak of the First World War in August 1914, but when Italy chose neutrality he was classified in Britain as an alien and may even have been at risk of internment. Eventually the British government permitted him to leave after first requisitioning his Company's sites. It was another slap in the face from his adopted Country, which he had made the centre of a worldwide revolution and manufacturing empire. Marconi simply went home.

CHAPTER FIVE

The War to End All Wars

The first two years after the completion of the Marconi factory at New Street saw the Company go from strength to strength. It even returned its first dividend to its more than patient shareholders in 1913. But the world was about to be torn apart.

At 5 a.m. on 30th July 1914, with the great naval review at Spithead just over, the 'first fleet', the British Royal Navy had just left Portland. It was urgently recalled by wireless telegraphy, and instructed not to disperse for manoeuvres as had been previously arranged. The wireless signals from the British Admiralty, sent via Marconi transmitting stations moved the Royal Navy's Grand Fleet to immediate battle stations throughout the world.

On 1st August 1914 the use of wireless was forbidden to all non-British ships sailing in territorial waters. On the following Sunday 2nd August the *London Gazette* issued a special notice that it had become:

> 'Expedient for the public service that His Majesty's Government should have control over the transmission of messages by wireless telegraphy.'

On 3rd August the Admiralty prohibited the use of wireless telegraphy on all merchant ships in territorial waters, providing for the dismantling of all wireless apparatus on merchant vessels in the territorial waters of the United Kingdom and Channel Islands. On the same day a second order decreed the immediate closure of all experimental wireless telegraphy stations in this country and arrangements were made for the equipment to be impounded. The communiqué asked for the co-operation of the public in order to secure information about any wireless station which may be observed to be kept up in contravention of this order.'

Radio was now at war; a vital asset that had to be safeguarded at all costs.

Across the North Sea, the German radio transmitter station at Nauen also sent out an ominous call to all German merchant shipping on the high seas to make for the nearest German ports, or, if too far away, for a neutral port.

On 2nd August 1914 German troops had entered France. On 4th August Belgium was invaded. At 11 p.m. the British ultimatum to Germany expired and the two countries were therefore, automatically, at war. Urgent wireless messages were sent to all units of the British Grand Fleet: *'Commence hostilities against Germany'*.

The world was now at war.

The Marconi transmitter at Poldhu Point broadcast the declaration of war across the world. The First World War bought huge and immediate changes for Marconi's and the Company was vital to the British war effort. As the world's leader in wireless communication it was inevitable that the whole organisation would be turned over to war work, including production, installation, training, research and development. For example every single message transmitted by the Germans throughout the war, in all some eighty million words, was intercepted by Marconi trained operators and passed on to the appropriate Government authorities.

Immediately on the outbreak of war the Marconi Works at Chelmsford was taken over by the Admiralty. The Clifden to Glace Bay transatlantic circuit was allowed to continue its function as a commercial station, but with interruptions and a change of wavelength to handle all naval traffic. The control of the wireless stations located at Caernarvon and Towyn passed into the hands of the General Post Office and later to the Admiralty, with Marconi's operating them for the Government. Stations for the interception of German wireless transmissions were hurriedly pressed into service at the Hall Street experimental station in Chelmsford, while the New Street factory itself was put under huge pressure to meet the demands of the armed services.
The Company's research and development departments went into overdrive and constant improvements were introduced. The delicate mechanisms employed in wireless telegraphy equipment had to be manufactured at a speed and in such

quantity as had never before been contemplated. In addition, there was a constant demand for simpler instruments for the instruction of wireless operations and students, including Morse keys, buzzers, telephones and headphones.

At the outbreak of war the Royal Navy was desperately short of trained radio operators. As merchant ships reached port, the civilian Marconi wireless operators were taken off and transferred to the Royal Navy. But this, while providing experienced men for the Fleet, in turn created a shortage in the Merchant Navy. The deficit was made all the more acute by the need to provide a much greater number of ships with wireless apparatus, as until 1914 only ships of more than 1,600 tons routinely carried wireless and these for the most part had only one operator. Not only were the big liners deprived of their Marconi operators, but ships between 1,600 to 3,000 tons which hitherto had not been fitted with wireless now found that it was a necessity. In addition, whereas before the war it was sufficient for one operator to be carried, now it was essential that there should be at least two operators to keep a continual listening watch.

Knowing that trained wireless operators would be in great demand, for some time Marconi's had been stimulating the interest of wireless amateurs by offering prizes for competitors in Morse code examinations and by making Morse practice sets readily available.

Immediately on the outbreak of war, the Admiralty took steps to secure the services of Marconi operators for all branches of the Service. In answer to the Government's call there came an army of lads and young men from all classes who had gone straight from school to the Marconi Marine Company, in whose offices they had been trained in Morse and the operation and maintenance of wireless apparatus. These at once volunteered their services to the Admiralty and the War Office, their places being taken by other lads clamouring to be trained as wireless operators. In due course the Company provided not only an army of operators, but also technical experts whose knowledge was unrivalled.

The Company's offices were open day and night, enrolling new recruits, instructing them on the art of wireless and examining them in Morse code. At the start of the war the Company undertook to find a further 2,000 operators to augment the 3,000 already serving on merchant ships. Purpose built classrooms at King's College and Birkbeck College were made available to ease the overload of trainees from Marconi House in London.

So great was the demand, that some of the pupils and enrolled scholars were as young as sixteen. The staff at Marconi House in London worked to the limits of their power and to the last ounce of their energy to meet the great emergency.

One thing was very certain, wireless was no longer the experimental toy of the Boer War; it was now a vital and indispensable tool for modern warfare.

In the early days of the conflict, a new radio based science known as wireless direction-finding had been developed. This new direction-finding (D/F) equipment used a 'soft' 'C' type thermionic valve, and was a modified version of an earlier Bellini-Tosi designed directional system. It had been developed just before the war by H.J. Round, a senior engineer with Marconi's working at New Street and Broomfield.

At the outbreak of war H.J. Round's work had quickly come to the notice of the War Office and he was soon seconded to Military Intelligence. He was ordered to provide an initial two D/F stations for service in France. This was speedily done and following their success, a large network, covering the entire Western Front, soon evolved to locate enemy positions on the ground.

These stations proved to be so successful that Round was instructed by the Admiralty to set up a second chain of stations in England, with the object of obtaining bearings on transmissions from enemy submarines. It was not long before similar networks were being built to maintain watch, not only for submarines but also for Zeppelins and German surface naval vessels. By 1916 the coastlines of Britain were covered by networks of Direction Finding wireless stations. Naval vessels were also fitted experimentally with D/F equipment that now included another one of H.J Round's inventions; a sophisticated error corrector.

In May 1916 the stations were monitoring transmissions from the German Navy that had been at anchor at Wilhelmshaven. On 30th May they reported a 1.5 degree change in the direction of the signals being picked up from the German fleet along with an increase in activity. The information was reported to the Admiralty who reasoned that the German fleet had put to sea. Accordingly the Admiralty ordered the British Fleet to put to sea to intercept the Germans and the following day the Battle of Jutland was fought. It was the largest sea battle of all time.

In it the British fleet lost seven ships and about 6,000 men, whilst the Germans only lost three ships (several others were seriously damaged) and around 2,500 men. While the British suffered greater losses, the battle of **Jutland is considered a strategic victory for the British.** While the British had not destroyed the German fleet and had actually lost more ships than their enemy, the Germans had retreated to harbour and at the end of the battle the British were in command of the area. Apart from two small and abortive operations the German High Seas Fleet was unwilling to risk another encounter with the British fleet and confined its activities to the Baltic Sea for the remainder of the war. Jutland thus ended the German challenge to British naval supremacy. For all his services during the war, Round was awarded the Military Cross but as a non-combatant he refused to accept it in uniform.

Just two days before the outbreak of World War I a Marconi engineer, Maurice Wright (later Director of Research), was experimenting with a new triode vacuum tube in a radio receiving circuit at the Marconi Laboratory at the Hall Street Works in Chelmsford, when he received German Naval wireless traffic.

Wright had joined the Marconi Company from university in 1912, and began work as an engineer on an improved method of detecting radio signals. Together with Captain H. J. Round, he succeeded in developing a vacuum receiver which made the interception of long-range communications possible for the first time.

By 1914, wireless telegraphy was in use by the practically all the world's military and naval forces. Signals officers and commanders in the field and at headquarters rarely took into account the possibility of interception, or deception. Unfortunately, the lack of really secure ciphers made wireless transmission risky. If intercepting cable telegraph traffic was simple, with wireless it was almost unavoidable. Messages were broadcast over the airwaves, and anybody could pick them up. Despite the lack of security, there was often no alternative to wireless, since it allowed governments to communicate with warships at sea and armies on the move.

On the first day of the war, a British ship dragged up Germany's transatlantic telegraph cables and cut them. From that time on, the Germans had to use radio links or telegrams sent through neutral nations, and both methods left them open to interception. As the Germans had advanced into Belgian and French territory where telegraph lines had been cut, it forced them to rely on wireless. The French and British, in contrast, only absolutely needed to use wireless to communicate with ships at sea.

Maurice Wright took the first batch of received German messages to the Marconi Works Manager, Andrew Gray, who was a personal friend of Captain Reggie Hall, the head of the Naval Intelligence Department. Hall was to become the dominant figure in British Intelligence during World War I and was responsible for cracking German ciphers from the famous Admiralty Room 40. He arranged for Wright to travel up to Liverpool Street Station on the footplate of a specially chartered locomotive. Hall realised the bonanza in his hands, and put Wright to work building a chain of wireless intercept stations for the Admiralty.

At the outbreak of the war, Britain had no formal code breaking operation. The British Admiralty's intelligence service began to intercept German wireless messages and quickly recognised the need for a formal cryptanalysis organization. Volunteer code breakers were found in the country's naval colleges. As early as November 1914 there had been a call in the British press for the use of private wireless stations to monitor for spy transmissions out of England and British radio operators were organised as the basis of the Royal Navy radio intercept service, feeding traffic to Admiralty Room 40 for cryptanalysis.

The intercept stations set up in this effort were known as the 'Y' stations. Marconi receiving stations, British Post Office stations and even an Admiralty 'police' station all provided intercepts to Hall's Room 40 codebreakers. These stations were soon joined by enthusiastic amateurs including Barrister Russell Clarke and Col. Richard Hippisley had been logging intercepts of German traffic at their amateur stations in London and Wales. New intercept stations soon went up on the coast. Soon practically all German naval wireless traffic found its way to Room 40.

The German high power long wave station at Norddeich provided much fodder for the codebreakers through the Y stations, which also soon turned to higher frequency interception as well. In 1915 these intercepts helped the British to win the naval battle at Dogger Bank, and played vital roles in later naval engagements.

The direction finding stations working under Round also provided intercepts to Room 40. The directional aerials tracked U-boats and Zeppelins as well as naval craft. The Y station intercepts also showed that the 1915 sinking of the *Lusitania* had the approval of the German high command, despite its continual denials. This along with the infamous intercepted 1917 'Zimmerman' Telegram, in which Germany promised Mexico it could have back the territory it lost in the Mexican American War, if it would join Germany against the United States was later instrumental in bringing America into the war. The leading history of the astonishing success of British intelligence in the First World War concludes: '[the] Y stations made it all possible.'

The First World War Radio intercept operators were largely unsung heroes; not of combat so much as of discipline. Much of their painstaking work had to remain secret. Today wireless interceptors have tuned their radio receivers all over the world for nearly a hundred years, often subject to all the risks of war, often in appalling conditions, often for impossibly long shifts, often without relief for weeks, striving for perfect copy of enemy traffic. After the First War wireless signals were rarely sent in plain language so the intercept operators could almost never understand the traffic they took down. They knew only that the signals came from an enemy and that they put their countrymen in deadly peril, and they did their duty.

During the First World War the British Army made considerable demands upon Marconi's. In August 1914, the Chief of the Marconi Training School at Broomfield, near Chelmsford, was seconded to the War Office and charged with the organisation of a large-scale Training School at the Crystal Palace in London for the instruction of officers and engineers of the Allied forces in the use of wireless in the field. At the same time a Field Station development section at the New Street Works was reformed as a separate department in order to meet the ever increasing needs of the armies overseas.

At the start of the war, the possibilities for wireless telegraphy on the battlefield were hardly recognised. Within the Army no separate wireless dedicated organisation existed, men from various units were detailed to attend to such wireless equipment as existed in addition to their normal duties. Senior officers were prone to look upon wireless as possibly a useful adjunct to visual and line signalling, but its main sphere of use was at first confined to communication between mobile cavalry and H.Q.

On the day the British Expeditionary Force landed in France, its total self-propelled mobile wireless force consisted of a single lorry fitted with one Marconi wireless transmitter and receiver. By the time of the first Battle of the Marne in September 1914, the force had expanded to ten units.

But by October 1915, the war of movement had ended. The advent of trench warfare brought to an end the limited role which wireless telegraphy had so far played. Instead an urgent new requirement arose. The main Corps H.Q needed to be fully informed, on a continuous basis about the situation on the front line. This was speedily organised by adopting a relay system consisting of a Morse code buzzer in the front line feeding to a 50 watt mobile spark transmitting set which in turn was in contact with a 120 watt spark set further back towards base. The ultimate link was a 1.5kW light motor set which was well within range of H.Q. Much of the output of the Marconi New Street Works in Chelmsford was at this time devoted to the provision of such sets. On 2nd January 1915, the importance of wireless in the British Army was recognised by the establishment of a Wireless Signal Company as a unit in its own right.

One of the key advances during the First World War was the use of wireless in the air. Hitherto a forward observer, often supported beneath a balloon, had to make sketches of what he saw over the enemy lines and then the information was dropped over the side and ferried to the British batteries as ranging aids. As the First World War progressed the possible advantages of communicating from the aircraft to the ground became obvious, especially for the immediate observation of the fall of shells and the reporting of troop movements.

The problem was that it was found to be impossible for the pilot or observer to tune his transmitter and operate a Morse key in an open cockpit, usually strapped to the top of his knee. It became clear that the ability to transmit speech was essential, not only for speed of command, but because the pilot of a single seat plane could not be expected to manoeuvre the aircraft and send Morse code at the same time. The other major handicap facing the introduction of wireless into aircraft was the very limited load capacity of the machines and the weight of the wireless apparatus then available. The installation of wireless in aircraft now required considerable experimentation, original thought and development. The early sets weighed over 80 lbs. and at first required some 250ft of aerial wire which had to be unwound by hand from a spool mounted on the fuselage alongside the observer's position. This in itself created several serious problems. In the event of attack it was impossible to reel the aerial wire in, so it had to be cut away. The wire also had a tendency to wrap itself around the aeroplane's control surfaces if its end weight twisted loose.

To cut down on engine interference, the ignition cables were screened with metal piping and sheeting, which not only added unwelcome weight, but also tended to make the aeroplanes engine even more unreliable. Another problem that beset early installation of wireless in aircraft was where to put the transmitter and battery. Extra items presented a problem for an already cramped cockpit, and the observer had to perch the transmitter on his knee, and keep the battery at his feet. All this equipment also left the observer and pilot virtually defenceless.

To solve these problems some of the best wireless engineers in the country were now rapidly given commissions in the Royal Flying Corps (R.F.C). The Experimental Marconi experimental section at Brooklands in Surrey had been formed in early 1911. It was hurriedly *'taken over'* by the RFC in 1914 and turned into a wireless training school for pilots and a research and development school for airborne wireless communication.

Over the next four years, the team successfully developed the science of airborne telephony using new valves and new techniques. The prototype sets were mainly built by the RFC's own workshops but for mass production it was Marconi's who stepped in to make it happen. Most of the RFC engineers joined Marconi's after the war and the skills and equipment they had developed were crucial in the rapid development of civilian air traffic and the birth of British broadcasting.

By 1916 all three armed services were depending heavily upon wireless. In the great Somme offensive of June 1916 it was often the sole means of communication between aircraft, artillery and infantry. Mobile wireless stations followed the infantry and the R.F.C, allowing divisions to communicate with each other.

By November 1918 the new RAF had some 600 aircraft fitted with 'Mark II choke controlled telephone sets', operating in conjunction with 1,000 ground stations and manned by over 18,000 wireless operators. Tragically there was an average of 400 casualties per year among RFC wireless operators, rising to nearly 500 lives lost during the months of May to November 1918.

Having entered the war as little more than a curiosity, wireless communication had in four short years become essential for the conduct of any military force, be it on land, sea or in the air.

Meanwhile, in Northern France the war dragged on in the mud and blood to its final and inevitable conclusion. It was a wireless message that had signalled the first act of war, now it was also the wireless that reported the end. On 11th November 1918, W.H. Chick, a special duty wireless operator at Marconi House in the Strand, intercepted the long-awaited message sent out from the Eiffel Tower station by Marshal Foch. The war was over.

Among the hundreds of wireless operators who were tuned to Eiffel Tower's familiar grunting tones was Guglielmo Marconi in his apartment in Rome. One can only guess at the emotions of the man who had in his youth foreseen wireless as a means of saving life and who, until his death, never ceased to look upon it as a potential means of promoting peace and understanding between the nations.

For Marconi personally the war proved a turning point in his life and career. In July 1914, at Buckingham Palace, King George V had awarded Guglielmo Marconi an honorary knighthood. This was an act of grateful recognition of his huge contribution to the new science of wireless but also, perhaps, a tacit apology for the harm done to his name by innuendoes of financial scandal just two years before.

Later that month, at Spithead, Marconi and his wife Bea lunched on one of the

Royal Navy battleships anchored for the annual review. Next morning the ships were gone, and within days, as a citizen of Italy, still neutral in spite of its 'triple alliance' with Austria and Germany, Marconi became an 'alien' and the subject of great suspicion. For a while, down by the Solent at Eaglehurst, his country house, he was even treated by locals as an enemy spy. He nonetheless remained there with his family until, in the words of his daughter Degna, 'sanity reasserted itself'.

Marconi left Britain and resumed travelling between the USA and Italy, where, as a non-combatant, he was elevated to the Senate in Rome in January 1915. He then sailed for the USA on the SS *Lusitania* to fight another legal battle in the on-going commercial war with Telefunken. In order to give evidence in a lawsuit against the German Company, Marconi sailed in April 1915 from Southampton to New York on the SS *Lusitania*.

On its return journey without him, on 7th May 1915 the ship was torpedoed by a German submarine and sank in just 18 minutes, killing 1,198 of the 1,959 people aboard, leaving 761 survivors. Some papers even speculated that the attack was an attempt to kill or even capture Marconi. The sinking turned public opinion in many countries against Germany and eventually contributed to America's entry into World War I. The *Lusitania* became an iconic symbol in military recruiting campaigns as to why the war was being fought.

Until 24th May, when Italy declared war against Germany, Marconi was involved in his American court case. He asked for an adjournment so that he could go home to serve his country and crossed the Atlantic safely incognito aboard the SS *St Paul*.

During the summer of 1915, Marconi was commissioned as a First Lieutenant of Engineers and put in charge of organising the Italian 'Army Wireless Service'. Then, as a Lieutenant Commander, he was asked to do a similar job for the Italian navy, which was suffering from the congestion of long-wave wireless traffic in the Mediterranean. To provide new channels for optical-range communication between ships, he went back to his earliest experiments at Villa Griffone, and re-examined the use of ultra-short waves.

In 1916, he asked a senior company engineer, C.S. Franklin, to design a small two-metre wavelength transmitter, and together they conducted sea trials the success of which led eventually to the development of the 'beam system' of short wave transmission during the 1920s.

Guglielmo Marconi then offered his contribution to the war effort by assisting with the installation of Marconi radio equipment on board ships and for the first time in aircraft. In reality it was a waste of arguably one of the world's most famous men, but none the less still good public relations.

After the Americans entered the war in April 1917, the Italian government sent Marconi on a number of goodwill diplomatic visits to the USA and Britain.

But he was in Rome early in the morning of 11th November when heard the end of hostilities announced by Marshal Foch in a wireless message transmitted from the Eiffel Tower.

During the war, 348 of his Company's operators had been killed, mainly at sea. But he continued to believe in wireless as a means of promoting peace. In 1919, Marconi served as an Italian 'plenipotentiary delegate' at the peace conference in Versailles.

It is no exaggeration to say that wireless had proved itself essential to the conduct of the war on both sides. On the allied side, at the lowest estimate, wireless played an important part in bringing about the enemy's defeat. Britain is an Island Nation dependent upon sources outside the United Kingdom for a large proportion of its food supplies and if the life-line across the seas had been cut, all the courage of this country's fighting men on land, on sea and in the air would not have saved us.

Wireless telegraphy placed in the hands of the British Navy and the Mercantile Marine an instrument which enabled them, albeit at great sacrifice, to maintain our supremacy at sea.

'They also serve' who, while not in the actual fighting line, are asked to devote their special talents and skill in other ways in the Nation's interest. All honour must be given then to those Marconi technicians, operators and instructors who, working day and night, supplied the Navy of Defence and the Navy of Supply with the apparatus and the gallant young men to operate it.

On 21st June 1922 Godfrey Isaacs unveiled a memorial plaque in Marconi House in the Strand to the 348 men of the Marconi Companies who had lost their lives in the war, the bulk of whom were sea-going wireless operators of the Marconi International Marine Communication Company. The plaque, which many years later was moved to the entrance hall of the new Marconi House building at the Chelmsford New Street Works and more recently to the Industrial Museum at Sandford Mill in Chelmsford bears the words:

'THEY DYING SO, LIVE.'

Wireless built at Marconi's New Street Works had gone to war and fought on land, at sea and in the air.......

But the world had paid a terrible price.

CHAPTER SIX

The Birth of British Broadcasting

'The Marconi Company during the 1920s was one of the most exciting places in the world for a scientist to work.'

Peter Wright

As the troops came marching home again, the giant Marconi Company returned to its peace time activities. During the war the Company had designed and supplied the armed forces on land, sea and air with the first generations of reliable wireless systems. It was now staffed by men who had survived the war, and whose baptism of fire had included working with radio equipment capable of transmitting speech. This was known in the jargon of the time as telephony, as opposed to telegraphy, which is the transmission of only the dots and dashes of Morse code.

Marconi's at New Street now returned to civilian management and began development of a new range of high power wireless telephony transmitters. These were known as 'panel sets', each capable of transmitting at different output powers, 0.25kW, 1.5kW, 3kW and 6kW. The Company also started design work on a range of high power valves to match the requirements of these new transmitter designs.

In late December 1919 Marconi's installed and began testing a 6kW telephony transmitter at the New Street Works. Operating under an experimental Post Office transmission licence, and using the radio call sign MZX, its sole purpose was again to investigate the properties and problems associated with long distance speech transmission. The transmitter was connected to a long 'T' shaped wire aerial that was suspended between the two massive, 450ft high masts. Normally the new trials would not have raised much comment as the Company was always testing new equipment at its Chelmsford sites. But this time something extraordinary occurred, almost by accident.

What happened next was to change the world.

Over the years Marconi's had developed a standard testing format for any new transmitting equipment. For these trials Marconi inland and shore stations throughout the country were told to listen for anything heard on 2,750 metres during specified periods. The times allocated for the early 1920 tests were between 11.00 and 11.30am and 20.00 and 20.30 GMT, for two weeks from 23rd February, except on Saturday nights and Sundays. The broadcasts were otherwise unannounced, so fresh from the success some earlier work undertaken at Marconi's Irish experimental station, Ditcham and Round considered that they had a free rein.

On the 15th January 1920 they started the first ever true speech 'broadcasts' in Britain by transmitting a programme of speech and gramophone music from the Marconi New Street Works. Included in this was what was to become Ditcham's regular 'news service', all transmitted from the New Street Research Department's Laboratory. The famous picture of Ditcham seated by the transmitter was taken here, but as with many Marconi publicity photographs of the time the background behind Ditcham and his transmitter was 'blacked' or 'whited' out, so as not to advertise how basic the engineers surroundings really were.

This first, historic accidental broadcast could well have gone unnoticed, but two hundred and fourteen appreciative reports soon arrived from amateurs and ship's operators alike who had listened in. The radio amateurs were enraptured to finally hear words and music on their radio sets, and they reported this in glowing terms to Marconi's. The Chelmsford station had been heard from Norway to Portugal, with regular reports over 1,000 miles and the greatest reported distance being 1,450 miles. A telegram from Madrid in January 1920 reported exceptional signal strength and quality. The engineering team realised that they had stumbled on something quite extraordinary. It was time to become more ambitious.

The 6kW transmitter was quickly replaced with one rated at 15kW input with MT4 and MR4 valves made in the Marconi-Osram works in Hammersmith. Ditcham, Round and another Marconi engineer, Mogridge were all involved with the design work and set about their task with great enthusiasm.

Now, for a brief period from 23rd February until the 6th March 1920 their continuing tests became a regular, scheduled series of 30 minute broadcast radio programmes. These were aired twice daily at 11am and 8pm and were designed from the outset to be a regular wireless telephony news service which would take up to 15 minutes, leaving time for three or four short musical items.

However, despite the enthusiasm generated by the Chelmsford radio 'events', Marconi's still believed at this time that the future of wireless telephony lay solely with commercial speech transmission and not entertainment. But it is to the Company's credit that it was always willing to allow individual engineers the time and budget to experiment and develop new ideas without complaining too much about costs. It was to do that just now and Ditcham and Round were given the free hand that their work deserved.

The new Chelmsford broadcast station came to life and additional entertainment was soon arranged, W.T. Ditcham became 'head cook and bottle washer', organising programmes, announcing news and music items and was ably supported by the Head of the Marconi Publicity Department, Arthur Burrows and Mr W. Petterigill. The regular programmes that Chelmsford now transmitted to the Nation during the summer of 1920 consisted of readings from newspapers, gramophone records, and for the first time, live musical performances.

The pioneering spirit of adventure was maintained, when the first ever paid artiste, Miss Winifred Sayer (later Mrs Collins) was invited to sing. Miss Sayer worked for the Hoffmann Manufacturing Company, just down the road (New Street) from Marconi's Chelmsford Works, but she sang as an amateur soprano with Edward Cooper and the 'Funnions' in the evenings.

She was paid the handsome fee of ten shillings nightly, for three half hour concerts from the new broadcast studio, in reality a packing shed next to the timber store, conveniently located across the road from the transmitter hut.

In reality the Marconi 'studio' had terrible acoustics. It was untidy, dark, cold and strewn with packing cases. There was no music for the lady's recital. Miss Sayer had to take her key from a tuning fork struck by Eddie Cooper, which meant that all the songs had to be short and simple, using an upturned telephone mouthpiece as a crude microphone. In addition, she had to contend with the hum of a fifteen kilowatt motor generator in the room next door. She was announced by Ditcham:

> 'MZX Calling. This evening for a change we have a vocalist; a lady vocalist too, you'll be glad to know, so I will now ask her to start on her first song. Will you start please?'

For three evenings Miss Sayer sang individually and also did a duet with Eddie Cooper. Her concert included the Edwardian ballad 'Absent'. Miss Sayer remembered that as she left she was informed by Godfrey Isaacs that she had helped to make history. But at the time she remembers that she was unimpressed by the new entertainment medium, and she called the whole affair a 'Punch and Judy show'.

Three months later she received a copy of Marconi's 'Souvenir of an Historical Achievement' booklet. In it were listed all the signal reports received by the Company including 145 amateur reports from all parts of Britain and Ireland. Land stations in Ghent, Lisbon and Norway all reported hearing the concerts. In addition there were 68 reports from ships at sea, eight being 1,000 miles or more from Chelmsford. She was staggered.

The Chelmsford station continued its concerts. The instrumentalists who provided these first live radio interludes were mainly drawn from the Marconi engineering staff. Mr G.W. White, an excellent musician and pianist organised the short musical passages and whenever possible played them on piano. Mr A.V.W. Beeton featured on oboe and Mr W. Higby played cornet.

Vocal numbers were also rendered by Mr Edward Cooper, a tenor, and Miss Sayers colleague in the 'Funnions' concert band. Eddie was normally employed in the Marconi Mounting Shop. Despite the very experimental nature of these telephony tests, and the sometimes variable quality of both transmission and quality, these unofficial musical interludes soon gained a large following.

The new concept of speech and music crackling over the airwaves into the front rooms of ordinary people was poised to revolutionise the world of entertainment. It was enthusiastically greeted by radio amateurs and newspapers throughout Europe. The transmissions from Chelmsford and Paris were joined by the new Dutch broadcast station, call sign PCGG, which started broadcasting special concerts for its English listeners on 29th April 1920.

It even appeared that the popularity of these experimental wireless stations had modified the somewhat reticent attitude of Marconi's toward the possibilities of using wireless telephony for entertainment purposes.

Remembering for a moment the experimental nature of these tests it is interesting to record that on the 6th March 1920, a telegram that still survives in the Marconi archives, arrived at Chelmsford. Mr B.T. Fisk, Managing Director of the Australian Marconi Company, reported that weak, but steady signals were heard in Melbourne on two of the Chelmsford broadcast schedules. Even though speech could not be clearly discerned, it appeared to all that the future of broadcasting looked very bright indeed.

Ditcham and Round's lively series of concerts and news programmes transmitted from the Marconi New Street Works had also not gone unnoticed by the established print media. The newspapers were soon to become very wary of this new mass communications medium, because it challenged their virtual news monopoly. But it was a newspaper that started the next phase of the story. This turning point in the history of British Broadcasting was brought about by Alfred Harmsworth, Lord Northcliffe. Northcliffe was the proprietor of the *Daily Mail* Newspaper group, and was an influential and successful newspaper owner who founded the *Daily Mail* in 1896. He had then revolutionised the British Press. His format was simple. Keep the stories less than 250 words, and include a murder every day. The stories also had to fit into his *personal idea* of what was newsworthy. The format worked and his newspapers which included *The Times* newspaper (since 1908) reached one in six of every British households.

It was Lord Northcliffe who now commissioned the first radio broadcast by a recognised professional artiste of international standing. He chose none other than the famous Australian Prima Donna, Dame Nellie Melba. He said that there was 'only one artist, the world's very best'.

By 1920, Melba was probably the most famous singer in the world. Between 1904 and 1926 she had made almost 200 recordings, and she had triumphed in the Opera houses of Europe and America. She had first sung in Covent Garden in 1888 and maintained her position in the golden age of opera for over 25 years, as her voice was remarkable for its even quality over a range of nearly three octaves, and for its pure, silvery timbre.

In her lifetime, Dame Nellie Melba achieved international recognition as a soprano, and enjoyed an unrivalled 'super-star' status within Australia, if not the rest of the world. But despite the angelic voice that Nellie Melba was admired for, she was also known for her demanding, temperamental and diva-like persona. It was this temperamental superstar that Lord Northcliffe wanted to bring to the new wireless audience, but getting the great lady to agree to sing was difficult. Reputedly when she was first approached the singer remained adamant that her voice was not a matter for experimentation by young wireless engineers and their 'magic playboxes'. It took all the persuasive talents that Lord Northcliffe could muster and a huge £1,000 fee, all paid for by *The Daily Mail* to get her to agree. Northcliffe had one other advantage. Melba, born on 19th May 1861 was now nearly 60 years old. By the time of the Chelmsford broadcasts she was approaching the end of her career and the promise of considerable newspaper publicity was something he knew Melba would not refuse.

Lord Northcliffe apart, it was really Arthur Burrows and Tom Clarke who should take the credit for the whole concert idea, and for then for making it happen. Burrows was Marconi's Publicity Manager and Tom Clarke was the News Editor of *The Daily Mail* and assistant to Northcliffe.

So Melba had agreed to sing for the wireless and a contract had been signed. But there was just one problem. There was nowhere suitable for her to do it. In 1920 it was totally impractical to move transmitting equipment and aerials to any of the great concert halls of Europe. The great Dame Nellie Melba would have to come to The Marconi Wireless Telegraph Company's New Street Factory in Chelmsford, Essex. Amazingly she agreed to leave the bright lights of the London stage for one night, and travel to a remote Essex factory. There would be no dressing room, no orchestra, no stage, no lights and in reality, no audience.

The 'Australian Nightingale' was booked to give her now historic thirty minute radio concert from the Chelmsford works, on 15th June 1920. But the concert announcement apparently sent the Marconi engineering team into a controlled panic, as they were given very little time to prepare.

It was thought that the 'best place' in the whole factory for the singer to perform was the first floor, Executive Directors and Senior Staff Dining Rooms at the front of the New Street Works and adjacent to New Street. With mahogany panelled walls, thick carpets and stylish period decor, the rooms were directly accessible from the main works entrance via an impressive curved staircase. This was considered a 'fitting' location for the great lady and would maintain the image of the Company. So the Executive Dining Rooms were set to become this country's first, professional radio broadcast studio.

The engineers' initial plan was to connect the transmitter, located at one corner of the works through to the new studio at the front of the building. This meant laying in a long, land-line cable right across the New Street Works, run it up the outside stairs, down the top floor corridor and into the Executive Dining Room. This would allow the singer to perform in pleasant surroundings, and keep her isolated from all the engineering and the engineers.

A huge amount of effort was also brought to bear on trying to improve the general quality of the microphone and transmitter circuits, but time was short. The cable was duly laid in, but almost at the last moment, during final testing it completely burnt out due to the high frequency current induced by initial test transmissions. The cable apparently actually burnt 'like a fuse' up the stairs. These tests also managed to destroy the modulating valves in the transmitter, causing great consternation among the already panic stricken engineering staff.

A hurried engineering meeting was called. The engineers were in favour of cancelling the broadcast, but this was considered impossible due to Dame Nellie Melba's concert schedule. Burrow's also pointed out, that to cancel now would cause massive damage, both to the reputation of Marconi's and to the cause of radio broadcasting in general.

So the decision was made to relocate the great lady's performance to the other side of the Marconi works actually a disused packing shed but very close to the transmitter hut. When the great day arrived and despite the valiant efforts of all concerned, the packing shed was still a gloomy place, its floors and walls bare of any decoration although a thick pile carpet had been placed on the floor, almost at the last moment. All the engineers could do now was wait.

W.T. Ditcham and the New Street Station MZX 15kW Transmitter

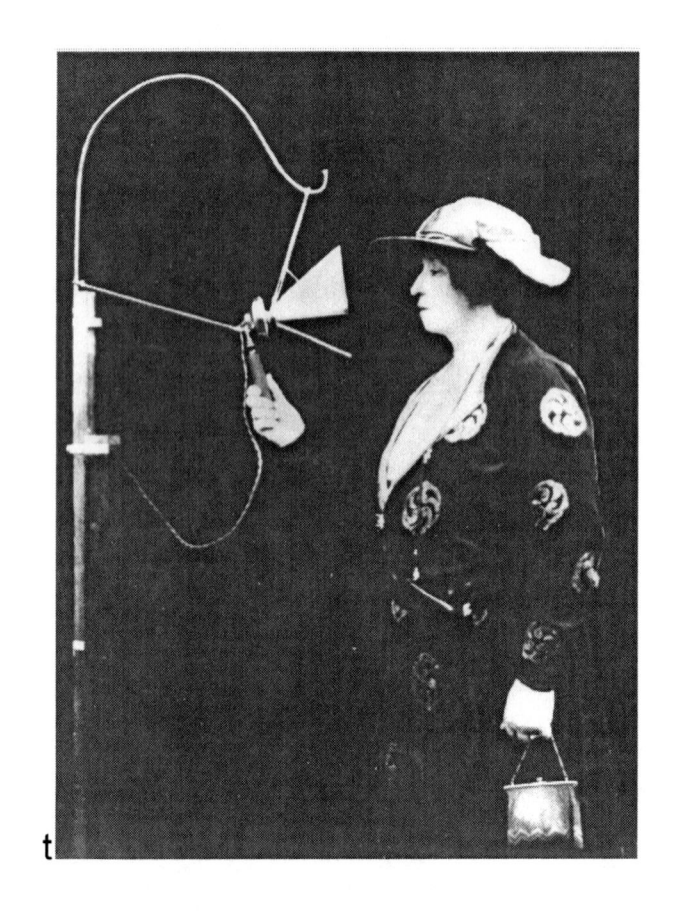

Dame Nellie Melba sings for New Street

On 15th July 1920, Dame Nellie Melba travelled to Chelmsford by train from London, wearing a black dress and a large white hat. At the station she was picked up by a chauffeur driven car and was then taken on a very long tour around Chelmsford, to eventually arrive at the front of the New Street Works. On a route advertised beforehand she was greeted by waving, cheering crowds before arriving at the main entrance, which in reality is just a few hundred metres from the train station.

Accompanying Dame Nellie was her son George Armstrong, together with his wife, and two of Melba's accompanists Frank St Ledger and Herman Bemberg, one of whose songs she was to perform. The party also included Arthur Burrows as her official escort, Godfrey Isaacs and his wife, Lord Northcliffe and his friend and colleague from the wartime propaganda bureau, Sir Stuart Campbell.

Dame Nellie Melba was first shown around the Works, including the transmitting equipment and the huge antenna masts by Arthur Burrows. Burrows remarked that from the wires at the top her voice would be carried far and wide. Her comment has become a piece of radio folklore:

'Young man' she exclaimed,

'If you think I am going to climb up there at my time of life then you are greatly mistaken.'

The early days of radio broadcasting were to be dominated by the enthusiasm and continual pressure generated by the growing band of radio amateurs, who always listened in and patiently 'tickled the cat's whisker'. Now, standing alongside Dame Nellie Melba, the Marconi engineers and the Chelmsford broadcast transmitter, Burrows knew that history was in the making. It was a testament to Burrow's amazing vision and determination that every experiment, trial and concert that built the foundations of British broadcasting over the next six years always had his guiding hand behind them.

On 15th July 1920 he was there again. It seemed that all was set for the great event, even the transmitter ran up without any problems. Before she could begin singing, Dame Nellie insisted that she must have her favourite pre-concert dinner of partly-cooked chicken, champagne and special unleavened white bread. This clause, or menu, was specially written into her contract, and had to be imported for it far outstripped the resources of the normal Company canteen. After her hearty meal in the Directors Dining Room, Dame Nellie was escorted across the works to the 'studio'. She passed no immediate comment on the room (to the relief of the engineers who had even hurriedly whitewashed parts of the walls) and seemed unworried by the Spartan surroundings. Her first glance at the studio floor brought her initial reaction, a firm kick at the new carpet. Dame Nellie announced to the assembled personages of the Company: 'First of all we'll get rid of this thing!'

Dame Nellie was not a woman to be ignored and the new carpet was hastily rolled to one side revealing the bare and un-swept stone warehouse floor. Despite this attempt to improve the room's acoustics, the engineers were still very worried as concert time approached. To fail now, with the whole world listening, might well endanger the cause of radio broadcasting for years to come. The newspapers had proclaimed that:

> 'At a quarter past seven this evening a great singer will hail the world by a long trill into space. Thousands of people on land and at sea are eagerly looking forward to hearing the glorious voice of the Australian nightingale Dame Nellie Melba swelling through space into their instruments.'

After an interval for photographs and other formalities the programme was arranged to start. The transmitters were run-up, but just as Dame Nellie was about to sing the last photographer in the transmitter room let off a flash bulb. The engineer on the switch panel saw the reflected flash, panicked and instantly pulled out all the switches, immediately powering down the entire apparatus.

While Dame Nellie watched patiently, St Ledger prepared his music, ready to accompany Melba on the piano and the painful process of transmitter warm up had to be gone through again. It was to be only a minor technical hitch. Ditcham was now happy.

On a perfect summer evening, it seemed that anybody who could pick up wireless waves throughout the country held their breath as Dame Nellie Melba stood in front of the microphone. Even this was something of a compromise, consisting of a telephone mouthpiece with a 'home-made' horn of cigar-box wood fastened to it during the final afternoon's testing by Marconi engineers, desperate to improve sound quality. The whole contraption was suspended from a modified hat rack by a length of elastic, but it has survived intact to this day; a powerful artefact from the very dawn of radio broadcasting.

Round later remembered that he had great difficulty placing Melba in the makeshift studio, especially as she told him that she knew all about it. The distances between microphones, singer, accompanists and the small grand piano were all critical in maintaining as high a quality as possible and Dame Nellie's position had been was marked on the floor. She eventually stood about a yard from the microphone. A hushed silence fell over the first-ever broadcast studio, and one of the Marconi engineers announced on air:

'Hallo, Hallo, Hallo!'

'Dame Nellie Melba, the Prima Donna, is going to sing for you, first in English, then Italian, then in French.'

The announcer then apologised for not having any control over the *atmospherics*. A chord was struck and at 7.10pm precisely listeners heard their first fleeting notes as Dame Nellie ran up and down the scale. This preliminary sound check brought a flurry of adjustments in the studio, the engineers swinging condensers and tapping meters.

It was, as the *Daily Mail* reported, a remarkable scene. Outside the 'studio' a large crowd had gathered, but the police requisitioned to keep order were not really required as everyone stood in almost reverential silence. Inside the room the group of VIPs had assembled. Dame Nellie called her first long silvery trill her 'hallo to the world', and the world seemed to be listening. All over the country wireless enthusiasts frantically tickled their cat's whisker crystal sets, desperately seeking a stronger signal. Headphones were clamped tighter and whole households lapsed into hushed silence. Dame Nellie took a deep breath, and began to sing:

'Punctually at a quarter past seven'

said a newspaper the next morning,

'The words of 'Home Sweet Home' swam into the receivers. Those who heard might have been members of the audience at the Albert Hall'.

This rendition was followed by Hermann Bemberg's 'Nymphes et Sylvains' (in French), Puccini's 'Addio' from 'La Boheme' (in Italian), and Bemberg's 'Chant Venitien'. Mr. Frank St Leger played the first two songs in the programme and Bemberg, the French composer, played the third.

History was being made from a disused packing shed on the edge of a huge factory. However, as Captain Round later recalled, the transmitter that had initially behaved very well during the concert, started to play up at the start of Melba's third song. Listening in anxiously on a wavemeter in the equipment room, Round watched in horror as mid way through the rendition, one of the transmitter valves started to fail. Then Chelmsford went off air.

Shouting instructions for the valve to be changed, he immediately rushed from the transmitter room to the shed where Melba was singing and waited for her to finish the last song. The repair had not taken long, but as far as the world was concerned the concert had abruptly ceased and her third song had been almost completely lost. Thinking quickly on his feet Captain Round called out:

'Madame Melba, the world is calling for more'

Dame Nellie Melba replied: 'Are they? Shall I go on singing?'

This was exactly what Round had hoped for, but whereas he had expected just one more song to make up for the partially lost one, in fact the good lady sang four more. While Round was 'pleading' for another song, W.T. Ditcham had corrected the fault and a minute or two later the notes of a piano again floated into listeners front rooms. After a brief pause for further adjustments to the transmitter, Dame Nellie gave a further encore, repeated 'Nymphes et Sylvains', and then sang the first stanza of 'God Save the King'. It was over. They had done it. Without fanfare, one of the Marconi engineers stepped to the microphone and simply said:

'Hallo, Hallo, We hope you have enjoyed hearing Melba sing, Good Night!'

The wireless sets of the Nation, indeed the world then lapsed into silence. The concert was over. The great lady had sung her considerable heart out and the assembled audience both inside and outside the studio spontaneously applauded. It was now up to the engineers to have relayed this to the audience with the clarity it deserved and hope that the partial failure had been compensated for by the extra songs. They knew that the audience was potentially huge and liable to be very critical, indeed nearly 600 new licences had been issued in the two months leading up to the broadcast.

Within days Marconi's was to receive its answer as enthusiastic letters from the four corners of the world poured into the Chelmsford office. Radio amateurs and ships' operators alike reported how they had listened in and sung along.

Every commercial station in the world had tuned in and the concert had been received with surprising clarity, and it was voted a great success. The amateur radio enthusiasts even amused themselves during the concert by listening in to the wireless station operators at the new Civilian air traffic control radio stations at Lymphe and Croydon discussing the performance between the songs.

Through all the publicity, acclaim and enthusiasm that the Chelmsford concerts generated, at least one person distinctly disapproved of the whole undertaking. The Postmaster General, The Rt. Hon Albert Illingsworth was definitely not amused. He promptly sent a strongly worded protest deploring the fact that a vital national service, such as wireless telegraphy, should be put to such frivolous uses.

Despite the fact that entertainment broadcasting was rapidly gaining favour with the general public, on 23rd November 1920 the Postmaster General spoke to the House of Commons. He announced that the experimental broadcasts from the Marconi Chelmsford Works were to be suspended on the grounds of 'interference with legitimate services' and for the time being no more trials would be permitted. Each experimental music programme from the Chelmsford, New Street site had to operate under a special Post Office permit, and there were to be no more permits. In reality there was also a political element to the closedown demand, as the Post Office was already seriously worried about its long held 'communications monopoly' in the British Isles.

One of the legitimate services that had been badly affected was the new Croydon Air Traffic Control System, itself built at the New Street Works. Typical was an article in the *Financier* newspaper on the 25th August 1920. It reported to its readers that a few days previously the pilot of a Vickers Vimy aeroplane was crossing the channel in thick fog and was desperately trying to obtain weather and landing reports from Lympne, but all he could hear was a Chelmsford musical evening. It also stated that:

> 'The opinion among airmen is practically united against a continuation of the 'concerts' given to the world at large by the Chelmsford wireless station.'

This view seemed to be echoed by the Navy and the Army, who stoutly maintained that any civilian broadcasting would hamper 'genuine experiments' and would not be 'in the best interests of Imperial Defence'. The critics of wireless broadcasting saw that the device was ideally equipped to be a servant of mankind, but were determined that it should never be considered as a toy to amuse children.

The Chelmsford broadcasts had also interfered with the Post Office's newly opened arc transmitting station at Leafield, near Oxford, that carried press and Foreign Office Morse transmissions on 12,200m to Cairo, India and America.

The Post Office also suggested that Chelmsford's broadcast concerts had more to do with publicity for the Company then true scientific development.

In particular, Lauritz Melchior's broadcast from Chelmsford had been heavily criticised for completely jamming all aircraft communications. Faced with all the complaints and problems, the Post Master General felt that he had no option but to close the Chelmsford station down completely. In also seems that Marconi's itself was not overly concerned. The concerts had been a huge drain on engineering resources and the Company saw little future profit in providing a free wireless telephony service for the press. On 9th August 1920, the Company gave orders that the 'special tests' should cease and that Chelmsford and Poldhu's standard tests were now to be limited to just 15 minutes. When it arrived, the Company complied immediately with the Post Master General's letter, and without any objection they transferred the high power transmitters from Chelmsford to Poldhu and over the water to the Clifden station in Ireland.

The Chelmsford MZX broadcast station lapsed into silence and the airwaves returned to their normal monotonous clatter of Morse code. But the seeds for the future of broadcasting had been sown at New Street. The Chelmsford engineers had focused the attention of both the press and the public on the possibilities of using wireless telephony as a means of bringing entertainment into the home.

It must be admitted that the Marconi licence was solely experimental, with no mention of broadcast entertainment programmes; but its cancellation at this crucial time in history denied British industry the chance to make strong bids for world markets in radio receivers and broadcast transmitters. This was a chance the

Americans were not slow to seize. Frank Conrad and the Westinghouse Company commenced broadcasting in America with station KDKA soon after Chelmsford started. This was quickly followed by hundreds of new stations, feeding news, records and concerts to an ever growing audience, who demanded radio sets. All this occurred just as British Broadcasting ground to a halt.

It must be remembered that in early 1920, the Post Office could never have foreseen that radio broadcasting would become such a powerful medium, nor that it would happen so quickly. Indeed it was felt by some observers that the amateur radio community actively encouraged telephony experimentation and broadcast experiments simply because Morse code communication was limited to those who could get their 'speed' high enough. To hear voices and even music in the headphones was on the other hand very exciting.

Regardless of methods or motives, under the strict control of the Post Office, the development of broadcasting in the United Kingdom was to follow an ordered and logical course, avoiding the total chaos that would soon beset the American airwaves. The sequence of events now turned full circle. The future of broadcasting in this country was returned to the small, but growing band of amateur radio enthusiasts, who had first appreciated the fascination of the broadcast spoken word. They were furious at the closedown of the Chelmsford station as it had been a vital reference signal operating on a precisely known wavelength and a declared power.

As a small compensation for the experimenting amateurs, after the complete wartime ban on radio amateur communications, experimental transmitting licences were finally granted again on 1st August 1920. Soon, more and more amateur telephony 'broadcast' stations began to appear on 1,000 metres and 180 metres, despite severe restrictions being placed upon their operation, including a maximum output power of only ten watts and operation limited to only two hours a day.

Lauritz Melchior sings for New Street

Marconi's Airborne Wireless Development Department, Writtle.
Home of 2MT

In reality, the genie was firmly out of the bottle. Once the Melba broadcast had been heard worldwide, radio broadcasting had become a sensational success. By March 1921 there were 150 transmitting licences and 4,000 receiving licences issued in this country, but some 1,700 requests or enquiries for experimental licences remained unprocessed. The amateur radio enthusiasts thought that it was high time for action.

The task of making it all happen, to build a transmitter and operate the station would be given to Marconi's Airborne Telephony Research Department, located in the small Essex village of Writtle. The content of the programmes would be determined by Arthur Burrows and his Publicity Department staff of 17 people working from his offices in the Strand.

The Writtle Research Department site had been established in anticipation of a new market in airborne telephony, when civil aviation got underway after the War. With the collapse of the German threat in November 1918 and the prospect of a return to more normal international relations, the growth of aviation as a means of civil transport was an obvious and inevitable development.

What was not so generally obvious at that stage, was the degree to which real progress in the development of organised public air transport was to depend upon a parallel evolution of airborne radio communications. It was some time before aircraft designers, and even airline operators accepted that wireless was essential for safety and organised air travel. In 1920 many still regarded the cumbersome and heavy wireless sets as being nothing more than an extravagant afterthought, and a not inconsiderable loss of pay load and even passenger space.

The young Writtle engineering team had already achieved some measure of success having within months of their creation designed and installed the 1.5kW transmitter for the Croydon Air Terminal. The equipment was built at New Street. The Croydon site used a double cage aerial 270 feet long and was at the forefront of the new age of civilian passenger transport. The only problem for Croydon and the Writtle engineering team was that the regular broadcasts from the huge Marconi Chelmsford transmitting station often caused severe interference.

Apart from providing one of the first ever air traffic control systems, Writtle also designed and tested the Marconi 'AD' series of radios, which by 1922 had been developed into the successful AD/1 Transmitter, AD/2 receiver, the AD/4 and AD/5 receivers, AD/6 transmitter, and later the AD/7 airborne transmitter/receiver. All these were also built at New Street.

So the new 'broadcasting job' landed unannounced on the desk of 'PPE' - one Captain Peter Pendleton Eckersley on the morning of 12th February 1922.

Almost by coincidence, the future of British radio broadcasting arrived in an ex-army wooden hut, parked unceremoniously on the edge of a large partly flooded Essex field. But the hut was staffed by the very ex RFC officers who had built the science of airborne speech transmission from nothing to meet the demands of the Royal Flying Corps. The years of research and development at the RFC's Brooklands, Joyce Green, Biggin Hill stations and since the war at Writtle, meant that this small group of engineers probably knew more about speech transmission over the radio than any engineering team on the planet.

At first it was thought that doing 'broadcasting' was just another job from Head Office, which would probably get in the way of the proper work. It was an inauspicious beginning, but with the irrepressible and hugely talented 'PPE' at the helm the small acorn would grow.

The Writtle station was to become the birthplace of broadcasting in Britain and over the next year the young engineers were to write the next chapter in the history of radio. Its call sign was simply 2MT.

Little did they know that the call sign and the hilarious antics of the small station behind it would change all their lives forever.

What they did next changed the world forever.

But that is a story for another place.......and another book.

CHAPTER SEVEN

Domestic Radios and the BBC

The instant success of 2MT at Writtle and the explosion of radio broadcasting in the United States quickly led many other companies to make applications for the few available transmitting licences in Britain. In March 1922, Metropolitan Vickers requested permission to start broadcasting in the North of England, and within two months, twenty more applications were received from other manufacturers. Parliament immediately came under great pressure from the pro-broadcasting lobby and agreed to set up a Committee to investigate the possibilities offered by the new broadcasting medium. Some kind of control or restriction was inevitable, but in reality few of the MPs knew or cared enough about broadcasting to think there might be a problem.

Shortly after 2MT began transmissions a rival to the Writtle broadcasts came on air as Marconi's had also been authorised to establish another experimental broadcast station. This time radio broadcasting came to the heart of the city, but initially only with a weak voice: 'This is 2LO Marconi House London calling.....' was the call, but it came at first at uncertain intervals and with an uncertain voice. The initial plan was to use the station for demonstrations to distinguished visitors, principally from foreign countries who might not want to make the trip to Chelmsford or the Writtle hut. The licence conditions for the operation of station 2LO contained a series of restrictions.

For a long time no musical sounds of any type could be transmitted, only speech was allowed using just 100 watts for one hour per day. At the end of every seven minutes' transmission there had to be three minutes' interval, during which the operator was compelled to listen on his wave-length for messages from Government stations telling them to cease transmitting. After a while the restriction on music was removed, and permission was obtained from time to time to use transmissions from this station to illustrate lectures being given in the London area on wireless telephony.

2LO was based in the attic training room above Marconi House on the Strand that had been opened on 25th March 1912. During the period of expansion in Chelmsford, Godfrey Isaacs had also decided that a new headquarters were necessary due to staff overcrowding in their existing buildings in London. As the New Gaiety restaurant was vacant, Marconi's made an offer to the owners, the London County Council, and a 99-year lease was agreed. The necessary structural alterations were completed in just ten weeks, and on 25th March 1912, Marconi House was officially opened. The building had a richly ornamented moulded plaster ceiling and its friezes were particularly beautiful. The panels were all mahogany and the grand staircase had stained glass windows at every level. On the landing were written Puck's words from *A Midsummer Night's Dream:* 'I'll put a girdle round about the earth in forty minutes'. Guglielmo Marconi had once retorted: 'I'll do it much quicker than that!'

The building's main lift was capable of carrying twelve passengers and to the right hand of the large waiting room was a telephone exchange containing three counters from which 'Marconigrams' could be dictated for transmission to all parts of the world.

2LO was a purely experimental station, but with the success of 2MT the Post Office came under pressure to allow some form of national broadcasting. By the end of May 1922 it had received 23 applications to start broadcasting and something had to be done.

On 18th May, the Post Office met representatives from 18 companies and asked them to come up with a cooperative scheme for broadcasting. Discussions went on for five months without any proposal. Each Company had its own interests and there was much conflict. The Government and the press complained at the delay and on 18th October proposals were finally put to the industry. The result was the setting up of the British Broadcasting Company, which would broadcast from eight transmitters, covering most of the larger areas of population. Opening capital and equipment was to be provided by the six largest companies: Marconi, GEC, BTH, Metropolitan Vickers, Western Electric and the Radio Communication Company.

Any British manufacturer or retailer could become a member by purchasing at least one £1 share. Listeners would buy a 10 shilling receiving licence and would pay two tariffs. The first was based upon the various components in the receiver and went to the BBC. The second was a levy of 12s.6d per valve holder which went to Marconi's as a royalty for use of its patents.

The privately owned BBC was the world's first national broadcasting organisation and was founded on 18th October 1922 as the British Broadcasting Company Ltd. The first transmission was on 14th November from station 2LO, located at Marconi House, London.

For the listening audience in the 1920s there were large numbers of radio kits on the market for the home constructor which soon became popular as they were a lot cheaper than ready built receivers. Although many different valve receivers were available from 1923, crystal sets remained the most popular receivers for several years. This was due to their low cost and freedom from the valve receivers' expensive high tension batteries and low tension accumulators which needed frequent re-charging. The early valves also had a relatively short life and so needed frequent replacement. Although crystal sets could only operate headphones, wireless exploded across the nation. By the summer of 1925, one and a half million receiving licences had been issued.

Valve receivers did however have the advantage of being able to operate a loudspeaker. Initially these were horn loudspeakers, but soon more modern types were developed which could be housed in the same cabinet as the receiver, to provide good sound quality. The early valve receivers were mostly of the tuned radio frequency (TRF) type, which often had an array of different knobs and switches on the front panel. Tuning-in a station required the operation of several controls, and to change waveband sometimes required the plugging-in of different tuning coils. Many receivers had a reaction control which adjusted the sensitivity and selectivity. If this was advanced too far, the receiver would burst into oscillation and act like a transmitter, so interfering with everyone else's receiver in the neighbourhood.

After World War I, the Marconi Company had begun producing non-industrial receivers, principally for the amateur market, at the Soho premises of The Marconi Scientific Instrument Company. In 1922, after its involvement in the birth of British broadcasting the Marconi Company formed the 'Marconiphone' department, to design, manufacture and sell domestic receiving equipment. This equipment complied with Post Office specifications and tests, and was therefore awarded the BBC authorisation stamp. The BBC/PMG Stamp was introduced on 1st November 1922 to fund the BBC by charging a royalty on all commercially manufactured wireless equipment. The new British Broadcasting Company's revenue depended heavily upon royalties levied on the wholesale price of all BBC/PMG-stamped receivers and accessories made by its member manufacturing companies.

In December 1923 the Marconiphone Company was formed to take over design, manufacture and retail of domestic receivers 'For those who lacked the necessary skills to construct their own'. Some Marconiphone Company sets were made at the Sterling Telephone Company (STC) Works at Dagenham. However, design and research for these domestic receivers still continued at Chelmsford. But the facilities and space in the Marconiphone department at the New Street Works soon proved totally inadequate to meet the huge demand. Mass production was moved to the Sterling Telephone Company works in Dagenham, the first sets being the Marconi Crystal Junior radio receiver, followed by the 'V1' radio set that used a single valve as the detector. The next set in the Marconiphone range was the classic Marconi 'V2', a two valve receiver with reaction applied. The 'V2' was for a long time the most sensitive receiving set available.

On 1st January 1927 a new non-commercial entity called the British Broadcasting Corporation was established under a Royal Charter and John Reith became its General Manager. Consequently on the last day of December 1926, the wireless manufacturing Companies ceased to be directly responsible for broadcasting in this country.

The radio industry was by now well established, and 1926 was to see the complete demise of the crystal set and the horn speaker, their replacements being the far more selective early valve sets. The 'new' BBC could boast ten main transmitting stations, ten relay stations and over two million listeners. From an original staff of four the new Corporation had grown to 552 in number.

In December 1929, the Marconiphone Company was sold to the Gramophone Company, along with the right to use the trademark 'Marconiphone' and the copyright signature 'G. Marconi' on domestic receivers.

Marconi's never re-entered the domestic radio market. In 1931 the Gramophone Company merged with the UK Columbia Gramophone Company to form Electric and Musical Industries Ltd (EMI). They produced domestic and radio receivers using the 'Marconiphone' trademark until 1956, when receivers were made by the British Radio Corporation, under licence. Domestic receivers bearing the Marconiphone trademark produced after 1929 had no connection with any of the Marconi Companies.

The new BBC in London and regional stations around the country gave birth to a new form of mass communication. 'Listening in' to the wireless throughout the United Kingdom, quickly became a social and cultural phenomenon.

It had been born and natured within the New Street Works.

CHAPTER EIGHT

Short Waves

Immediately after his military service in World War One, Marconi had turned his attention to short-wave directional transmission. In 1919 Guglielmo Marconi bought the 700-ton steam yacht *Rowenska,* a luxury boat that had once belonged to the Archduke of Austria. He renamed her *Elettra* and equipped it as his floating laboratory and second home. On the *Elettra* he could work in solitude, spend holidays with his children and entertain his many 'high society' friends across the world including royal and titled grandees, tycoons and Hollywood film stars. In 1924, he was made a Marquess by King Victor Emmanuel III.

In Britain his Company pioneered radio broadcasting and the new BBC but the Italian pioneer was not involved. On board the *Elettra* Marconi conducted a long series of experiments on short wave communication which he stated needed 'one-tenth of the power required before'. His research work dramatically increased the transmission ranges possible and he was convinced that this was the future of long distance wireless communication.

First mooted in 1906, Marconi's grand plan to link the British Empire by a network of wireless communication stations had run like a disruptive thread through the Company's history for the next twenty years. Long before the First World War, Britain had considered changing from a system reliant on cables to a wireless system for all communications across the Empire.

In March 1910 Godfrey Isaacs had submitted to the Colonial Office an imaginative plan to link the British Empire by a network of wireless stations, by means of which different portions of the Empire would all be put into communication with one another at greatly reduced rates. He proposed to begin with a series of eighteen stations, taking in Egypt, India, Malaya, China, Australia and Africa. At this time he was asking for licences for the stations and for the support of the Government in obtaining licences from self-governing Colonies.

The Company, he said, was prepared to erect, maintain and operate stations entirely at its own expense providing the government would suspend development of the long-wave stations until Marconi's short-wave beam system had passed its trials. The announcement by Marconi that short-wave communication seemed to be more promising threw the government into disarray. The great advantages of the new system were that smaller aerials and reflectors could be used and much less power was needed to achieve the same results. In addition the capital cost of beam stations was a tenth of that of cable and the operating costs were also lower. This 'Imperial Wireless Scheme', it was argued, would also be of great strategic advantage, providing ships of the Royal Navy with a global means of communication unhampered by vulnerable landlines and submarine cables. The British government hesitated, for fear of handing Marconi's a monopoly, but under pressure from the Imperial Defence Committee, agreed to examine the possibilities.

This plan was first considered by a Standing Committee called the Cables (Landing Rights) Committee, which in March 1911 reported against the suggestion that Marconi's should own and operate the stations, but recommended that a state-owned system connecting the Empire was desirable and that Marconi's should be approached to erect it. In June the Imperial Conference endorsed this suggestion and a committee was formed, with the Postmaster General in the chair, to negotiate with the Company on behalf of the Post Office. Negotiations began in the autumn of 1911, and on 7th March 1912 a tender was signed between the Post Office and the Company which provided for the erection of the first six stations.

But there was considerable disagreement about the idea and then a huge scandal erupted over possible illegal Company share dealings within the Government of the day. Although the so-called *Marconi Scandal* was resolved, further plans for the grand communication scheme were then interrupted by the First World War. It would be another seven years until a formal contract was signed with Marconi's.

The delay was probably fortuitous as now Marconi knew that the use of shortwave beams promised a greater volume of traffic at much higher speeds. But despite the advances in wireless made during the war, Marconi's vision was still derided as 'amateur science' by a Royal Commission in 1922. One member even concluded that radio was 'a finished art.'

Marconi re-issued his challenge. He offered to build, free of charge, any link across the world, provided the government would suspend long-wave development until the beam system had passed its trials, and provided they would adopt it if the trials were successful. The government agreed and specified the toughest contract they could devise. The agreement with the Company was: 'that as the contractor, it should build an initial beam station for communication with Canada. If this proved successful, the system could be extended to communicate with South Africa, India and Australia'.

The Government also asked for a link from Grimsby to Sydney, Australia, and demanded that it operate at 250 words a minute over a twelve-hour period during the trials without using more than 20kW. Finally they demanded that the circuit be operational within twelve months.

These were incredible requirement specifications. Radio was still in its infancy and little was known about generating high power at stable short wave frequencies. The project would have been impossible without the commitment of the Marconi technical team including H.J. Round, C.S. Franklin and Maurice Wright. The first operational site for the new global communication scheme was the Canadian Beam station which was opened at midnight on 5/6th October 1926.

The technical requirements for the Grimsby-to-Sydney link had astonished the rest of the radio communications industry. But the Tetney station near Grimsby in Lincolnshire established communication with Australia at 6 a.m. on 8th April 1927 and India at midnight on 5/6th September of the same year. It worked twelve hours a day for seven days at 350 words a minute and was undoubtedly one of the great technical and engineering achievements of this century.

The Tetney station was paired with its receiver station at Winthorpe, just north of Skegness. Tetney was the 'pilot' station for all the following Beam stations. It provided two services: to Sydenham near Melbourne in Australia, and to Dhond near Poona in India. The Australian service used the callsign GBH and operated on a single wavelength of 25.096 metres (11954 kc/s). Of the two routes, the west-bound was longer (approx. 12,000 miles) than the east-bound (approx 9000 miles) and the direction was switched to whichever had the longest part of its path in darkness- west in the morning, east in the evening. The Indian service operated on 35 metres (8.6 Mc/s) for the long path, and 19m (15.7 Mc/s) for the short path.

The next beam station was at Dorchester, paired with Somerton near Bridgwater in Somerset as the receiver station. Initially the station operated two services, to New York and to South America. These were supplemented by services to Japan and Egypt by 1928.

The Ongar Radio Transmitting Station occupied a site of 730 acres at North Weald in West Essex adjacent to the late 19th Century North Weald Redoubt. It was originally built in 1920 and operated by Marconi's before becoming part of the Beam system. The station operated two services, to Paris with callsign GLS and to Berne with callsign GLQ.

This was the beginning of the Imperial Wireless Chain - a revolution in worldwide communication.

The typical beam stations used all Marconi equipment designed and built at New Street. The transmitters were type SWB-1, short for 'Short Wave Beam', and generally pronounced 'swab' amongst engineers. They were specially designed for this service and used CAT2 or CAT3 power triodes in their final stages which were the first valves to incorporate copper to glass seals; the transmitters achieved an output of 11kW. These valves were cooled by paraffin but this was changed to water cooling at the start of WW2 to reduce the risk of fire in the event of the station being bombed.

For the first time the chain home transmitters used demountable valves allowing rapid replacement of damaged or burnt out valves. The vacuum in the valves was maintained by a continuously running pump, a system that was to be critical for the successful operation of the cavity magnetron valves in later radar equipment.

The beam antenna system was designed by C.S. Franklin and was a curtain array of stacked dipoles for each service, suspended between 280 ft towers. Elements were cut for a number of different frequencies to permit day/night working were provided, and the elements could be switched to reverse the direction of the beam. At Tetney, the towers were spaced 650ft apart, and the total array was nearly a mile long. Twin feeders of copper tube were used which resulted in a very low loss feed line.

All the stations were extremely successful. The Australian Government had been guaranteed an average traffic capacity of 20,000 words per day. In practice the capacity proved to be over three times this amount. In a single week in December 1927 the total number of words carried by all the Beam Stations was at a rate which would total 34,840,000 words per year.

Marconi's great ambition to provide a world-wide system of wireless communication was realised. Marconi's New Street Work's had built the equipment that now joined the four corners of the world.

Alongside development of the beam system Marconi's were still working for the BBC. On 27th July 1925 the remaining rural areas of Britain came within crystal set range, when the 25kW (later increased to 30kW) station at Daventry started transmission, relaying the London BBC programme.

The station had begun life a year earlier on 21st July 1924 as an experimental station, (hence the call sign 5XX) designed, built and operating on the long wave (9200-600 metres, 1500-500 kHz) instead of the medium wave from the Chelmsford Marconi Works. The original Chelmsford transmitter design had generated 15kW and gave over 40 amps in the large drainpipe aerials at the New Street site. Immediately the station came 'on-air', many listeners from the areas of previously very poor reception wrote to both Marconi's and the BBC, praising the new station's quality and clarity.

The new Warwickshire transmitter site on Borough Hill, Daventry was some 600 feet above sea level. Daventry had been chosen because it was the point of maximum contact with the land mass of England and Wales. It was accompanied by increased power and two aerial masts of 800 and 500 feet in height. The call sign 5XX was transferred from Chelmsford to Daventry, where its wide range, helped by 25kW output, and reliable reception on a slightly lower frequency of 1554m would make it famous. Chelmsford took the call sign 5GB for future operations and testing.

The various stations of the BBC could now be heard by 85% of the country and could also be heard throughout Europe. The 29th October 1930 edition of *Wireless World* reported that the long wave 5XX was still the only British station of any real entertainment value. The medium wave broadcast stations were often subject to fading and Morse code interference especially in fringe areas.

Alongside the development of the beam system and 5XX the first experimental BBC short-wave broadcasts to the Empire were sent out from the New Street Works in November 1927. Using special aerials supported by the tall New Street masts, New Street pioneered the overseas world services of the BBC. From 1932 the BBC Empire Service (now the BBC World Service) was also moved to and was broadcast from the Daventry station. The radio announcement of 'Daventry calling' made the station name well-known across the world.

Prior to the construction of the beam stations, the majority of world message traffic was carried by the traditional cable system but within months of the opening of the beam service it had captured a large proportion of the business. The seed that Marconi had sown in 1901 had grown into a mighty oak. Within six months the beam system had taken 65% of Eastern Telegraph traffic and more than 50% of Pacific Cable traffic. Cablegrams cost 6d at the ordinary word rate and 2 shillings for full rate while by beam radio the full rate was 4d. The cable operators began to feel the effects of the competition and their revenues collapsed.

For Marconi's the expenditure involved in developing the beam system had been a considerable drain on the Company's resources, but Marconi's 'Beam System' proved so successful that the government intervened to prevent a monopoly developing. A committee was set up under the chairmanship of Sir John Gilmour which led to The Imperial Wireless and Cable Conference at which the Companies and Governments were represented. The remit of the conference was to review the situation which had arisen and to make recommendations with a view to a common communications policy being adopted by the various governments.

In 1928 a report was produced which recommended that the cable and wireless resources of the British Empire be merged into one system.

As a result a new Company was formed under the name 'Imperial and International Communications Ltd.' This name was changed in June 1934 to Cable & Wireless Ltd. which is the name still used today. It was recommended that the Company would have a revenue target of £1,865,000 and that it would lease the Post Office beam stations for £250,000 per annum for 25 years until 1953.

In September 1929 all the Marconi Beam stations were transferred to the new single operating Company which also controlled 164,400 nautical miles of submarine cable, 13 cable ships, 253 cable and wireless stations and offices and a stock of new cable worth £824,000. The new Company owned the entire system until the passing of the Commonwealth Telegraphs Act, 1949, whereby the United Kingdom radio services of the Post Office and Cable and Wireless Ltd. became integrated on 1st April 1950.

The Marconi S.W.B.8 HF transmitter was originally produced in the late 1930s as a shipboard transmitter, having an all-brass carcass construction for corrosion prevention. Many units were used by all branches of the Allied forces in World War II including the Australian Army, and after the war about a dozen units were purchased by Australian Department of Civil Aviation. Given the DCA stores identification Y5-6081, these units were commonly referred to as 'Swab 8s'. They were used on point-to-point radio teletype circuits linking Sydney, Darwin, Perth, Cocos Island, and to overseas stations.

Very reliable and robust, they were a good example of the best quality professional transmitter available at the time. They remained in Departmental service from the mid-1950s until the mid-1980s. High power output (up to 3.5 kW) was required to overcome atmospheric noise at the lower frequencies used. In 1938, SWB-10 transmitters were introduced. These had an output of 25kW, using water-cooled final stage valves, with components fully enclosed and interlocked.

The Marconi SWB18 shortwave transmitter was developed primarily for the BBC external services and went into service during the early war years. It was designed to work between 13.5 and 80 metres with a carrier output of 100 kW up to 10 MHz and 75 kW above. A very fast change of wavelength was achieved by use of an ingenious design of interchangeable 'trucks' carrying pre-set high frequency tuning circuits that could be wheeled in and out.

The SWB series of transmitters were all designed, developed, manufactured and tested at New Street and probably rank amongst the Company's finest and long lived equipment types. They represented everything that was good about British engineering. They looked like they had been designed and built to last a hundred years, constructed from heavy gauge brass and copper sheet. Despite the speed at which the equipment was engineered, examples of the SWB1 type transmitter of 1924 vintage were still handling daily beam traffic in 1962.

For Guglielmo Marconi the beam system represented the crowning achievement in his career, but by the end of the 1920s he visited England very infrequently. In Italy he was worried about *Il Duce*, the Italian dictator Benito Mussolini. As President of the Italian Royal College of Science, Marconi automatically became a member of Mussolini's Fascist Party; privately, he thought it was rowdy and opportunistic.

Concern for his health grew after Marconi suffered two attacks of angina, but he would not slow down. Early in 1931, having set up a short-wave station for the Vatican, he supervised the Pope's first broadcast to Catholics worldwide. In the same year, on the 30th anniversary of his first transatlantic signal, his own voice circled the world. In 1933, with his wife Maria Cristina, he circled it in person.

Between 1931 and 1935, returning to his earliest research Marconi experimented with microwaves. At sea, aboard the *Elettra* in 1934, he used this research to develop blind navigation by radio beacons essential for the development of civilian air transportation, but also the first step toward a new and critical technology that would later be called radar.

Marconi died in Rome on 20th July 1937 at age 63 at his home in Via Condotti following a series of heart attacks. Italy held a state funeral for him. In a gesture that was unique among all the tributes that followed, wireless stations closed down and transmitters all over the world fell silent for two minutes. For a brief moment the ether was as quiet as it had been before Marconi was born.

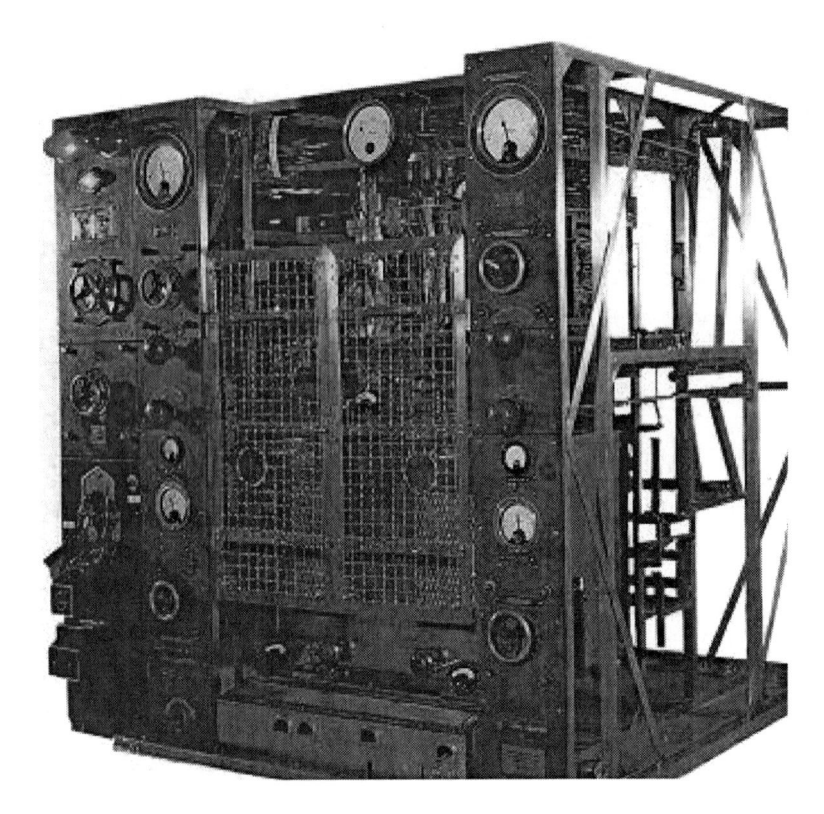

SWB 1 Transmitter

SWB 8 Transmitter

SWB 11 Transmitter

Daventry 5XX Transmitter

Marconi's loss was deeply felt in the Company, but in reality it activites and growth had long outstripped the control of its founder. Alongside the huge beam transmitters Marconi's were working on many other fronts.

Due to the ever increasing danger of aerial collisions in bad weather on 19th November 1933 the first UK Air Traffic Control Service was established at Croydon in London, based on the Marconi system that had first been designed and installed by the Writtle broadcasting station pioneers over ten years earlier.

The new system laid out an irregularly shaped wireless based Control Zone based on geographical features that came into operation when visibility dropped to below 1,000 yards or cloud was below 1000ft. Standard routes were also defined over south east England, mainly following railway lines and using the 'right hand traffic' rules to avoid collisions.

Alongside the ATCS work, Marconi's was also working on advances in telephone exchanges, a business that continued with the Company until its very last days .

But Marconi's New Street Works also had its social side and the Company always looked after its workforce. The three main Chelmsford factories each had their own sports field. Hoffmann's was in Rainsford Road, Crompton's was at the lower end of Wood Street and Marconi's was in Beehive Lane. Sports days, especially inter works competitions between the three factories were eagerly anticipated with a full programme of running, hurdling cycling and field events. The Tug of War match was probably the most popular staged between the biggest men each works could find.

There was always an impressive array of prizes, cups, shields, plaques and a dazzling selection of cut glass vases, biscuit barrels, rose bowls, canteens of cutlery and luxury items all donated by the Directors of the participating Companies.

The three factories also participated in the annual Hospital Carnival organised by Mr R.G. Morrish, Secretary of the Chelmsford and Essex Hospital in London Road. The many multi-coloured Floats, bands and walking entrants all marshalled in Rectory Lane and then preceded along Broomfield Road, Duke Street and through the town, the Mayor taking the salute from the steps of the Shire Hall.

Percy George Warren demonstrating one of the first PBX telephone systems developed at New Street, 1929. Photo from his granddaughter, Caroline Moth.

Alongside the equipment and technology Marconi's was a Social Company. New Street Sports-day, 1932

From High Street and then Moulsham Street the procession turned right through Writtle Road and down London Road eventually arriving back at the recreation ground. Then, traditionally the contestants on each factory's Floats wrecked their opponents display.

But back in the main New Street Works, alongside the BBC and Beam transmitters, ATCS and telephone systems there was another techonogy brewing within the Company. Marconi's last microwave experiments had sown the seeds of a new technology, at first simply known as RDF – 'range and direction finding'.

It used the same initials as the established Radio Direction Finding systems, this time to conceal its ranging capability. It would save the Nation and help win another World War.

We now know it as radar.

CHAPTER NINE

RADAR

As early as 1916 Marconi and C.S Franklin, working in Italy had turned their attention to short waves (about two metres) and found in the course of their experiments that these were being reflected by obstacles in the path of the signals. Marconi did not pursue the speculation, not least because the development of the short-wave beam communication system took precedence, but around 1930 Franklin wrote:

> 'Shadows and reflections from objects when using such (ultra short) waves had been observed for some years. I was interested in forming a wireless picture using centimetre or millimetre waves in a manner analogous to the optical picture obtained by a camera. It can be done ...but owing to changing conditions I was never able to get the work organised.'

The earliest mention of the concept of radar within the Company came on 20th June 1922 when Guglielmo Marconi addressed a joint meeting in New York of the American Institute of Electrical Engineers and Institute of Radio Engineers receiving the latter's Medal of Honour on the same occasion. He said:

> 'In some of my tests I have noticed the effects and deflection of these waves by metallic objects miles away. It seems to me that it should be possible to design apparatus by means of which a ship could radiate or project a divergent beam of these rays in any desired direction, which rays, if coming across a metallic object, such as another steamer or ship, would be reflected back to a receiver on the sending ship and thereby immediately reveal the presence and bearing of another ship in fog or thick weather. One further great advantage of such an

arrangement would be that it would be able to give warning of the presence and bearing of ships even should these ships be unprovided with any kind of radio'.

What we now know as radar, started development in the United Kingdom in the mid-1930s after Marconi had demonstrated the principles of blind navigation by wireless beacons in the early 1930s.

With the foundation of the Third Reich in 1933 Adolph Hitler began to construct a huge air force as part of his rearmament plan. The British Air Ministry recognised this force as a serious threat and in October 1934 Dr H.E. Wimperis, the Director of Scientific Research at the Air Ministry and his colleague Mr A.P.Rowe set out to review the ways in which science might be utilised to defend the realm. The outcome was a Committee of Research on Air Defence under Mr H.T. Tizard but early experiments with acoustic and infrared detection of aircraft soon proved technically unworkable,

In January 1935 Wimperis consulted Robert Watson-Watt, the Superintendent of the Radio Department of the National Physical Laboratory to investigate the possibility of using some form of 'death rays' or high power electromagnetic radiation to incapacitate the pilot, disable aircraft motors, or detonate the bombs of an approaching airplane. Following a study, he concluded in his report to the committee in January 1935, that the so called 'death ray' was impractical; but that detection of aircraft appeared feasible.

On the morning of 26th February 1935, Robert Watson-Watt parked his van is a field near Daventry in Northamptonshire for the first radar experiments in Britain. With his assistant, Arnold Frederick Wilkins he successfully detected reflections at a range of about eight miles from a metal clad Handley Page Heyford bomber as it flew through the beam from the empire service shortwave transmitter of the Marconi built radio station at Daventry (Borough Hill) located a few miles away.

They used a standard radio receiver of a type in everyday use for research and measuring upper atmosphere reflections of short waves installed in a trailer. The histroic field was at Stowe Nine Churches (just off the A5 about three miles south of Weedon Bec in the Daventry area). The interference picked up from the aircraft allowed its approximate navigational position to be estimated.

By 13th May a special laboratory had been built at Orford Ness on the Suffolk coast. In June the Tizard Committee saw reliable detection of aircraft at ranges of up to 40 miles. Watson-Watt's team demonstrated the capabilities of a working prototype and then patented the device (British Patent GB593017).

The pace of development was expanded on 1st September 1936 when Watson-Watt became Superintendent of a new establishment under the British Air Ministry; the Bawdsey Research Station located in Bawdsey Manor, near Felixstowe in Suffolk. Here aerial towers 250 high built by Marconi's, allowed the tracking of aircraft flying at over 1,500 feet and 75 miles away. Further work resulted in the design and installation of aircraft detection and tracking stations called 'Chain Home' (CH) along the East and South coasts of England in time for the outbreak of World War II in 1939.

Transmission at any one CH station required four 360-foot-high masts, 180 feet apart, between which the antenna wires were strung. The returned signal was not received by the same antenna, but on four separate 240-foot-high masts. The transmitter aerials 'floodlighted' the airspace in front of them with pulses of radio energy which, when reflected from an aircraft, was picked up by the receiver aerial. The range of the 'echo' was directly measured on the face of the cathode ray tube and the position of the target could be ascertained through triangulation from other stations with a radio direction-finding instrument called a goniometer, essentially an instrument that measures an angle.

The original chain of RDF stations (Radio Direction Finding – the term 'radar' was not adopted until 1943), were built from Southampton to the Tyne. The first was finished at Bawdsey in 1936, which also served as a radar training school and was handed over to the RAF in May 1937. In 1937 Marconi's had acquired the expertise to fulfil Government orders for transmitter aerials for the Chain Home stations. By July 1937 the first three stations were ready and the associated system was put to the test. The results were encouraging and an immediate order by the government to commission an additional 17 stations was given, resulting in a chain of fixed radar towers along the east and south coast of England.

By September 1938 over £2 million (today over £112 million) had been spent on the new secret weapon and Britain was covered by a security blanket capable of tracking individual aircraft in all weathers up to 15,000 feet and over a hundred miles away.

At the outbreak of war in September 1939, CH had eighteen stations covering the eastern and half of the southern coast of Britain. They all reported into one 'Filter Room' which became Britain's first Air Defence Radar Network, vital during the Battle of Britain. By the end of the war over 50 had been built.

Alongside the radar system and antenna development, Marconi's worked on the 'heart' of the radar system, a new air cooled valve called the magnetron. The development and production of this device was a vital Marconi war effort. The cavity magnetron is a high-powered vacuum tube that generates microwaves using the interaction of a stream of electrons with a magnetic field. The 'resonant' cavity magnetron variant of the earlier magnetron tube was invented by John Randall and Harry Boot in 1940 at the University of Birmingham, England working under Admiralty direction, Their pioneering work resulted in the resonant cavity principle used in the magnetron which was capable of generating very powerful radio signals of many kilowatts on minutely short wave-lengths (50 cm to 150 cm) and its invention and development made centimetre radar a practical proposition.

When put into production by Marconi's it became the heart of the secret battle fought out between the laboratories of Britain and Germany.

Magnetron production began in August 1940, but the challenges of the cutting edge technology meant that by the end of the year only 20 a week were being produced. It was auch a great novelty that the first units were passed round the Great Baddow Laboratories like a new scientific toy. Marconi's also started supplying the General Electric Company with the Magnetron 'blocks' which were made under conditions of great secrecy. Just as the Magnetron is the heart of the radar transmitter, so is the 'block' the framework of the Magnetron, and the whole basis of its operation. This component involves a number of machine-tool operations carried to fine limits, and in a material not specially well suited for mechanical processing.

Even so, towards the peak of the output, the Marconi Works were producing hundreds of 'blocks' a week for different types of Magnetron. As time went on, new types of Magnetron were introduced, and the Company was firmly established as one of the three main supplies. The great secrecy which was observed in connection with the Magnetron was such that even scrap 'blocks' had to be accounted for, and a serial number stamped on every one of them, all duly recorded.

With other manufacture underway the Great Baddow laboratory moved to Waterhouse Lane to begin volume production and 9,000 sq feet of floor space was dedicated to production in a former Marconi van garage. Soon several hundred cavity magnetrons were being manufactured each month and by 1945 a peak of 2,500 a month was reached

But at the outbreak of the Second World War many of the Marconi Company's most gifted scientists and engineers were seconded to Government Research Establishments for work on radar and allied subjects. At this time, most research and development of radar was carried out within Government units, although substantial manufacturing work was contracted to Marconi at New Street.

CHAPTER TEN

The Birth of Television

'Gentlemen, you have now invented the biggest time-waster of all time. Use it well.'

Isaac Shoenberg,
Head of the Marconi-EMI TV development team.

As sound broadcasting became an everyday event at the BBC, engineers started work on the science of broadcasting pictures. The BBC Television service began the world's first regular public service of high-definition television programmes on 2nd November 1936 from the Alexandra Palace station in North London, using an aerial and transmitters designed and made at New Street.

The Marconi-EMI television system was chosen by a Television Committee to provide the regular BBC service after a short trial with the rival Baird system.

Early TV pioneer John Logie Baird's system relied on an antiquated method of mechanical scanning to form the picture. By the time of the trials an all-electronic television system was possible, developed since 1934 by the Marconi-EMI Research Team led by Sir Isaac Shoenberg. The Baird system was cumbersome, inflexible and expensive and was probably at the limit of its development. What is probably the most remarkable thing about the whole trial process was that Baird even managed to persuade the BBC to transmit his signals, which were of such poor quality that it was doubtful if it had much entertainment value.

The Marconi-EMI system was flexible and capable of much further improvement. EMI talked about the flicker-free nature of their system with its new 'EMItron' camera and the fact that it was capable of capturing truly instantaneous images.

The Marconi-EMI system could offer three cameras, two mobile and one fixed with the main EMI studio divided into three, with a different act performing in each section one after the other, the cameras and lights moving backwards down the studio as the show progressed.

In the other camp Baird's system was fixed, low definition, prone to flicker and his actors and presenter needed special high contrast make up. The system also suffered a minute long delay, as it relied on Baird 'Intermediate Film Technique' that required baths of hot cyanide to develop and also limited the system to around 20 minutes per program.

> 'No whizzing discs, no mirror drums: silence, lightness, portability. It showed the way things were going. It was quite easy to see, even then, that the Baird system couldn't eventually lead anywhere...'
>
> ***Douglas Birkinshaw***
> BBC Engineer In Charge of Television

However the important thing about the new art of television was that it proved that it was possible to transmit moving pictures of live events over a public network. These could be viewed 'live' by anyone with the appropriate apparatus, which at first was a relatively simple device that could be attached to a normal radio receiver.

When the decision was made to start a full BBC Television Service, the first thing the new Department had to do was make a technical choice. The location of London's TV station had to be high up, as VHF waves need line-of-sight reception, even in Band I as these transmissions would be, around 45-49 MHz. The decision was made to use 30,000 square feet at one end of the decaying Victorian entertainment complex, Alexandra Palace, in Wood Green, North London. The BBC was to provide the 215-foot mast and antennas (one for sound, one for vision), and the Marconi manufactured sound transmitter. The impressive 'turnstile' antennas were over 600 feet above sea level, and got out extremely well: the intended range was just 25 miles, but reception was in fact achieved in Manchester and occasionally on the Continent.

The construction of the television complex took place during the summer of 1935, and was completed in mid-1936. It was to include two sets of television studios and control rooms, initially one for the Baird system and the other for Marconi/EMI system, who were in competition with each other for the right to broadcast on a regular basis.

The world's first high-definition television service was formally launched by the BBC on 2nd November 1936 at 3 p.m. with Leslie Mitchell announcing. The 240 line Baird system and the electronic system produced by Marconi-EMI using 405 lines were used on alternate weeks. Marconi sourced their camera from RCA who they part owned. Baird on the other hand started using Farnsworth's camera under license for his broadcasts.

Television used a standard 3 kW Marconi broadcast transmitter for the sound, but a new type of 'vision' transmitter was needed to cope with this emerging technology and the combined efforts of Marconi and EMI produced a remarkable design. Marconi were responsible for the radio frequency sections, while EMI developed the modulator. In 1936 the new high power transmitter at 45 MHz with a bandwidth of some 3 MHz was on the very cutting edge of technology and it was to the great credit of the engineering teams that the equipment was still in service in 1956. It had a rated peak white output power of 17kW and the aerial system produced an effective radiated power of 34 kW; the maximum power allowed at the time.

The station employed separate sound and vision aerials. It was Marconi engineer C.S Franklin, one of Guglielmo's Marconi's first engineers, who after a long and successful career with the Company designed both the sound and vision antenna, just before he retired in 1937.

A newcomer to the TV station would immediately recognise the familiar style of a Marconi HF transmitter, complete with the standard control desk of the period, but there the similarity ended. At purpose-built wireless stations the machinery was always installed in a dedicated area well away from the transmitters, in order to divorce them from the inevitable noise, but Alexandra Palace was somewhat different. The building had not been designed with the idea of housing a large and powerful transmitting station. The accommodation consisted mainly of a number of very large rooms, one of which housed the vision transmitter and machinery while the sound transmitter was located in another room.

As the operator sat at the control desk, facing the transmitter at one end of the room, and behind were the lines of motor generators. In addition, the main EHT was produced from a 50kW, 500Hz motor alternator which screamed away at the other end of the room, supplying a mercury arc rectifier. The overall noise made such things as monitoring the sound programme or answering the telephone almost impossible. There was an oscilloscope which displayed the sound carrier so at least one could tell if the transmitter was working.

At the start of television, with two sets of almost everything provided, the new service was, in reality a spectacularly wasteful duplication of effort. After the event, it became evident that there had been political motives behind this complex and expensive compromise protecting the government and the Post Office from criticism that would have resulted from excluding Baird, who was considered by many to be the pioneer television broadcaster. It also seems that bankruptcy of the Baird Company, already in severe financial difficulties, was highly likely and the Government wanted to avoid any blame for the company's collapse if they excluded Baird.

After just three months of trials the Baird transmissions were discontinued when it was concluded that the 405-line Marconi/EMI system was far superior to the 240-line Baird system. The final Baird transmission went out on 30th January 1937. From then on, with a break during World War Two, and right up until 1981, television programme production continued from Alexandra Palace.

The effective range of the new Television service covered approximately 25% of the UK population, and was much greater than had been expected. In the beginning there were very few television receivers, (possibly only 400) rising to around 1,000 within a year. By 1939 when the service was closed down at the onset of World War II, there were an estimated 23,000 sets in use by the public.

> 'I believe viewers would rather see an actual scene of a rush hour at Oxford Circus directly transmitted to them than the latest in film musicals costing £100,000'
>
> **Gerald Cock**, Director of Television,
> in the Radio Times of 23rd October 1936

Many people believed that television sets, then selling at a price equivalent to a family car, itself a luxury beyond most people's reach, would remain too expensive for mass sales. Cinema companies were also unsure about television as in the 1930s the cinema was a hugely popular medium, with picture houses in every town, and many people attending more than once a week. However, they did see potential for showing large screen news programmes which could be relayed instantaneously, something impossible to do with film.

Despite the limits of the system's range, around 30-50 miles, which excluded most people living outside London, and the high price of sets, plus the limited programmes available (about an hour or so daily), electronic television proved successful. Most of the major radio industry names started to manufacture commercial television sets, but almost all used the cathode ray tube type of display which was an important part of the EMI system.

Television was still an expensive novelty. Estimates of the numbers of receiving sets sold in the period 1936 to 1939 vary from several hundred to a couple of thousand. From 1938 it looked as though television was here to stay and plans were made to extend coverage to the rest of the country, and increase the number and variety of programmes. The radio industry knew that sales of wireless sets were approaching saturation and that the new technology of television would be a challenge, but it was potentially a huge new market.

Even at the 1939 Radiolympia Exhibition, the major trade show in August where next year's models were on display, every major company had TV sets or combined radio and television's on their stands. Catalogues showed many new 1940 models under development, most of which were probably never made.

After the first successful broadcast, Isaac Shoenberg is reputed to have (mis) quoted Hugh Latimer, Bishop of Worcester in 1555, by telling the Ally Pally team that: 'We have this day lit a candle that will never be put out.'

But the outbreak of War was to put an abrupt brake on the whole concept of television. On 1st September 1939, the BBC ceased television broadcasting to prevent it being used as a target navigation beacon by German bombers. Although the instruction had gone out to close down at noon, it seems that an outside broadcast from 'Radiolympia' overran and this was followed by a Mickey Mouse cartoon which began at 12.05 p.m. It would seem that those on duty wanted to delay the closedown for as long as possible. 'Mickey's Gala Premiere' was thus the last television programme broadcast in Britain for seven years. There was no closing announcement, just a test card for a quarter of an hour, then blank screens and static. The transmitter was switched off at 12.35 p.m. The studio doors were locked and the staff moved on.

It can be questioned why the BBC poured such huge sums of money into the new system, which in reality could only service a small part of the British population. In addition Marconi-EMI brought huge resources to bear, working not only on new camera designs but also on higher vacuums, new techniques in glass ware, electronics and on improving the performance of receiving-end cathode ray tubes, developing screens larger than 7in and a true black-and-white display. In addition television had required the development of new volume manufacturing techniques and facilities for the associated technologies.

But all these technologies and techniques were to prove critical for the future war effort as the science of very high frequency radio waves and high vacuum cathode ray displays had another use, which was top secret; radar. Many of the engineers involved in the development of television were at once put to work on what would become an enormously important defence science that would save the lives of thousands of servicemen and civilians. In the early 1930s, as Hitler rose to power in Germany, someone in the Government, BBC or Marconi's had recognised that war was inevitable.

Television studios were closed and silent, although the Alexandra Palace transmitter was used to misdirect enemy bombers using a radio guidance system when it came back on air as part of 'Operation Domino', a very successful counter-measure to the *Y-Gerat* beams used to guide German aircraft to their targets over British cities.

From the beginning of the war the Luftwaffe had used systems employing radio beams to enable bombers to find their targets. The first, 'Knickebein', was relatively easy to jam. The second, X-apparatus or 'Wotan I', proved more difficult and before effective countermeasures were in place it was used for targeting the terrible raid on Coventry, amongst other towns and cities. By the end of 1940 this too had been beaten by the British intelligence services. A new, more complex system was therefore anticipated and sure enough, it duly arrived.

Early in 1941 the Luftwaffe began to use a new bombing and navigational guidance system called Y-apparatus or 'Wotan II'. It employed a transmitting station on the Cherbourg peninsula that broadcast a signal down a narrow beam on a particular frequency. The beam was to be aimed directly at the target for that night. The lead aircraft in the formation flying along the beam received the signal and rebroadcast it on a slightly different frequency. This was picked up at the Wotan II station and the position of the aircraft determined by the time delay. Thus the bomber could be given a signal at precisely the right moment to release its bombs.

As luck would have it, the frequencies used were within those covered by the Alexandra Palace transmitter. The message went out to call back BBC engineers who could fire up the transmitter again and they were found in the nick of time. Thanks to some excellent intelligence, countermeasures were in place ready for the first night the system was used operationally by the Germans. The signal rebroadcast from the aircraft was picked up at Swain's Lane receiving station and sent along cables running through various tubes and tunnels to Alexander Palace. It was converted back to the same frequency used by the Wotan II transmitter and broadcast by the powerful television transmitter. Thus the aircraft received the same signal twice. The combination of the two signals modified the phase shift and thus the apparent transit delay. Initially, the signal was even re-transmitted at a low power, not powerful enough for the Germans to realise what was happening, but enough to spoil the accuracy of the system. Over subsequent nights, the transmitter power was gradually increased eventually totally jamming the system.

The development of Television was on hold for the duration of the War and in the homes of those who had been fortunate enough to have a receiver, sets were put away or used only as radios and would remain useless until 1946. The BBC television service re-opened in 1946, from Alexandra Palace by showing the same Mickey Mouse cartoon used on close down. There cannot have been many viewers of the early transmissions, as the number of working TV sets within range would have been less than in 1939.

Many pre-war sets would have been faulty after many years of disuse, and indeed there is evidence to show that some were returned to the factory to be overhauled, or in some cases completely rebuilt. A television customer was a valued one, and the scale of production was low enough that companies such as EMI or RGD were willing to bend over backwards to keep their loyalty. A 1939 combined TV/Radio set in the higher price range would have cost about the same as a terraced house in many towns and the buyers had received a raw deal as before the war their new equipment had provided only a few weeks or even hours of viewing. Then followed six years with a useless cabinet stuck in the room and probably a blown fuse and a blank screen when transmissions resumed.

The climate in manufacturing was also not conducive to the rebirth of the service. Materials were in short supply and even wood for cabinets was scarce. Some pre-war designs were put back into production, notably the Cossor 900 and Marconi VT50A, but these were badly out-dated thanks to the development of new valves and techniques during the War. The Government wanted British industry to concentrate on export markets, but there was no export market for TV sets, and barely enough resources to make radios which could be exported.

But television survived and grew. In 1964 the BBC introduced its second television channel, this time using the 625-line system. This gave even better definition than the 405 system. The new line standards agreed were those used by many other broadcasting authorities, so enabling the international exchange of programs to take place.

Alexander Palace Television Transmitter

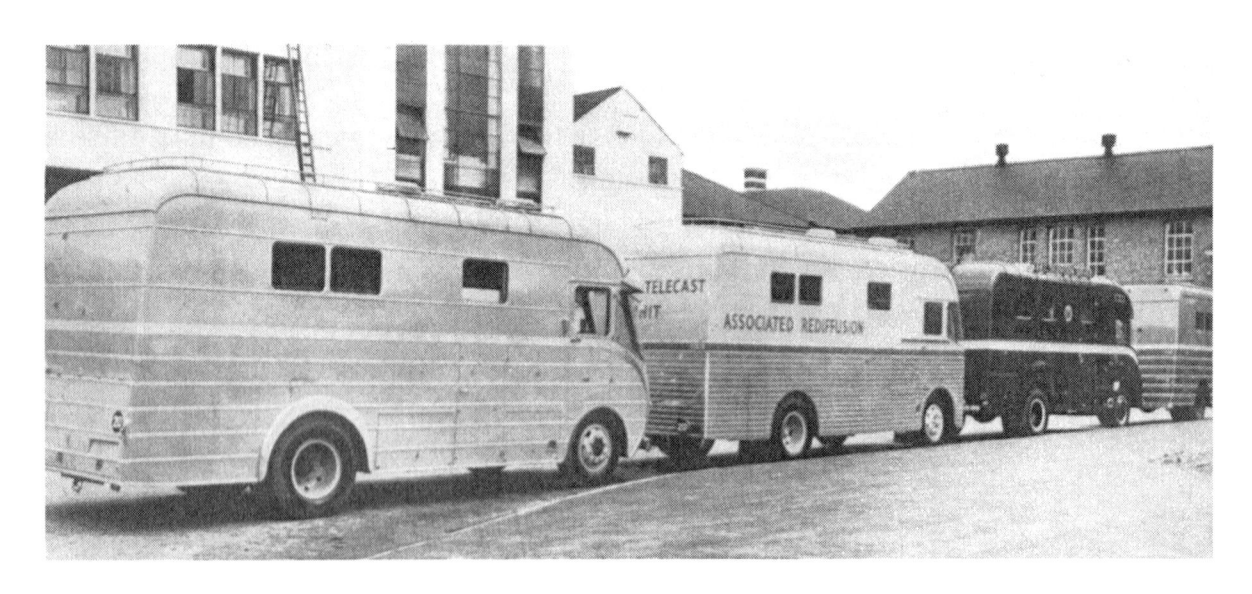

Outside Broadcast vehicles, New Street Main Yard, 1952

Today there are thousands of television transmitters across the world. Of course they now operating in colour first adopted in the 1960s. Many of the transmitters, cameras, outside broadcast units and aerials were supplied from the New Street factory.

Television is a relatively new medium, but already it has gone through many changes which have come about mainly through technical development. It has also changed its position in society, from an experimental novelty, through an amusement for the wealthy, to probably the most powerful medium in Western culture.

But technology marches on at an ever increasing pace and television may already have passed the peak of that power and could yet become merged with other digital media in the next few decades.

CHAPTER ELEVEN

The Second World War

Marconi New Street in Wartime Camouflage

Not long after the introduction of television, war with Germany was declared on 3rd September 1939. All the Marconi Companies were again called on to support the three armed Services and the Merchant Marine. Alongside the secret development of radar other 'crash' programmes had started as early as 1938 to provide our bombers, fighters, ground stations and warships with essential radio equipment.

Even the Marconi Research Laboratories at Great Baddow, completed in March 1939, were taken over by the Air Ministry.

In the dedication to 'Marconi - A War Record', Admiral H.W. Grant, Chairman of Marconi's Wireless Telegraph Co., Ltd. wrote:

'This War Record is offered as a tribute to the men and women of the Three Fighting Forces, the Merchant Navy, and the Fishing Fleets by some who served them in workshop, office and laboratory, where were forged those invisible weapons that overcame Space and gave eyes to those who fought evil in darkness…'.

One day, early in 1938, the Chairman sent for the New Street Works Manager and said:

'I want you to consider all measures for the maintenance of power – un-limited and assured power – against the emergency of war. And I want you to make all provision for emergency communications, and for proper shelter accommodation against the same contingency.'

By the end of 1938, the Works Engineer was able to report that the New Street works now had an electrical plant independent of the grid able to provide 80 per cent of its normal load. He was also able to report that the flourishing Company Fire Brigade was at a double strength and doing weekly fire drills.

In May 1939, when war was seen to be inevitable, a control centre into which some two thousand tons of concrete were poured took shape in the Works yard. It remains there, to this day, with its three-feet-six inch walls and seven-feet-six inch roof.
(I wonder if the future developers of the site realise that it is under there – I remember that the demolition company who took down part of the works to make way for Eastwood house (on a fixed price) were 'surprised' by the air raid shelter, alarmed by the huge concrete aerial base and devastated when a JCB actually fell into and disappeared in the underground cellars from the original brewery that stood where the top car park is now. Tim Wander -Editor)

The Munich crisis was the signal for the digging of a defence system round the works, a rush of A.R.P. (Air Raid Precaution) measures, including look out positions on top of Marconi House and the camouflaging of the Works. An additional sub-station assured a power supply should the main sub-station become a casualty.

By 3rd September 1939, when the voice of the Prime Minister Neville Chamberlain, broadcast news of the declaration of war, the Marconi Company was already on a war footing with production fully ready for the coming onslaught.

Marconi's had received its first service orders early in 1938. They were Admiralty contracts for transmitters and receivers of all types, for both big ships and small craft. On 6th October 1939, a member of the Designs Department in the New Street Works hurried into the Shops and thrust a paper into the hands of the Works Manager.

'This is a red-hot job for the Air Ministry,' he announced, 'Now, go to it.'

Scribbled in pencil on the paper – a Works Note – were these words: Install S.W.B.8 short wave transmitter equipment in coaches to design instructions.

The Works Manager thus faced with the urgencies of another 'crash' job, returned a laconic 'O.K.' and began at once picking a team of men suited to carry the work out at maximum speed. While the foreman was still hand-picking his team, a string of luxury motor coaches were being driven into the yard on the south side of the camouflaged works. They represented the fruits of a London-wide search by an official of the Company and an Air Ministry officer armed with plenary power of requisition.

Not many minutes later a group of men came into the yard from the Works, swung into the coaches and began ripping out their upholstered and chromium-plated interiors. The 'crash' job was under way. Recalling the occasion at a later date, the foreman of the job said:

'Normal practice is for a job to be carried out from drawings prepared for us by the Designs Department, and these indicate clearly every detail.

'But there was no time for that. So our instructions were just word-of-mouth as we went along. And it was 'Do this' and 'Do that', and 'Tear that out' and 'Shove this over' and 'Knock that away'. And we went to it.'

The precise nature of the task was the conversion of the ten luxury motor-coaches into two heavy mobile field stations, complete with transmitter, power plant, central telegraph office, receiver and accommodation coach. It was a job that, working normal hours, at normal speed, from full drawings, might have been completed with credit to all concerned in four to five months. Working a twenty-four hour day, with some men putting in a twenty-four-hour shift now and then, and one man sustaining a continuous forty-eight-hour shift, the first conversion was completed by 12th October; the second five days later.

On that day, a group of exhausted, sleep-drunk men watched the R.A.F. drivers take over those transformed vehicles that combined, with their suggestion of power and efficiency, the promise of speed. And someone said jestingly: 'Ah, there goes the Blue Train!' And the name stuck. This was the first of the 380 war-time commercial-vehicle conversions made at Chelmsford, each of which was equipped with S.W.B.8 or S.W.B.11 units. They were mostly used by the Army, and often on gun sites for jamming the enemy. It was an output equivalent to one complete conversion every nineteen hours throughout the War. One more 'train' went out of the Works yard. It was made up of seven Bedford-Scammell vehicles converted into a mobile transmitting and receiving station. It was christened 'The Golden Arrow'.

At the start of the war Marconi's General Manager, who has grown up with the Company remarked:

'Our plant is not a factory though things are made in it. It's a model shop. Most work in most factories means repetition. It involves the manufacture of standardised types and models adapted to conveyor-belt production – the belt moves and each worker merely performs a single, simple operation at a set speed, day by day, month after month.'

'It's not so with the production of modern wireless and radar apparatus, where the work ranges from the making of large S.W.B.8 transmitters, with chassis of brass weighing from ten to twelve hundredweight, to small components wrought to a thousandth of an inch of accuracy by micrometer measurement.

'Such work can be done only by skilled and experienced craftsmen, and the man is more important than the machine; and neither can move until the scientist and technician have finished their jobs. For it is their work, after all, that is the soul of the organism.'

In 1939, the Manager of the Development Department was called to Portsmouth by the Admiralty. As the Company had been producing naval equipment in large quantities for a long time, there was nothing strange about this call. The Manager and his Chief Assistant were received in a bleak hut in which two senior naval officers awaited them.

'What we want to know,' the Scientific Naval Officer enquired, 'is whether you can adapt your equipment to our special naval requirements?'

He was told: 'Yes.'

'And how long will it take?'

'It's an eighteen months programme.'

'Well, we must have it in a month – ONE MONTH.'

This was just one of the first of those jobs that came at intervals throughout the War, The first model was produced in six weeks and three days; exactly a fortnight after the outbreak of war. Marconi wireless apparatus and personnel had already been employed by both the Army and the Royal Navy since 1899. The claim may be added that since 1902 the Company has also continuously serviced the Merchant Navy, for in that year both *Campania* and *Lucania* were first equipped with Marconi wireless sets.

The equipment called for had been designed originally for strictly limited production, and the problem set the Company's engineers was adapting it to quantity production. This was achieved by centralisation of design of the three classes of receiver, of which the CR 100 (as B.28) became standard in the Royal Navy, and was also adopted, to a considerable extent by the other Services.

By this method of standardization a phenomenal stepping-up of production was achieved. Whereas in 1938, marine-class receivers were being produced sometimes by the few hundred, sometimes by the dozen only, the total war-period output of all receivers, including high-speed diversity units as used by Cable and Wireless and the Post Office topped the ten thousand mark.

In the New Street workshops five hundred girls made 'simple things by simple processes, and complex things by processes that had been developed over many years'. The 'shops' were full of turning wheels, rotating lathes, whining drills and percussion hammers, making up the counterpoint against which poured forth a warm flood of broadcast music from green speakers slung overhead.

The Works never slept. The great Instrument Shop topped by electric lights, set in parallel, consisted of long lines of gear-littered work-benches. These, in turn, seem to converge towards a distant point at the far end of the Shop where lights and machines and moving works fused into a dim kaleidoscope that is the Instrument Shop.

Both Shops, Instrument and Tool (where the machines were so delicate that they must be kept at a constant temperature) were housed under a single roof, and, between them, they absorbed around thousand operatives, men and youths, women and girls, ranging in age from sixteen to seventy.

In June 1939, the Company employed 1,200 operatives; by June 1940 it employed 1,900. The increase was steady. By August 1944, 6,000 people were employed. In 1938, the Company worked a five-day week. Throughout the War a seven-day week was worked. During that entire period, the Works were never shut, save partially and temporarily, as the result of direct enemy action.

Throughout the war Marconi's and the New Street Works did not only produce cutting edge equipment and technologies. One of its essential outputs were basic but vital screening components for all wireless apparatus fitted into aircraft that had to be completely screened from the intense high-frequency currents, magnetos, leads and sparkling-plugs electrical noise generated by its engines. During the war Marconi's manufactured some 72,150 screening sets in all.

Another key component produced was the 'Stabilivolt' and production started early on in the War. The Stabilvolt had been developed at Great Baddow research and was a special valve used to control the voltage of radio receivers when the supply was taken from a mobile source liable to large or rapid changes in voltage. Before the war the valve had been imported from Germany, but now a new source was vital to the war effort..

The work on this unit was new and there were many technical and production difficulties to overcome. Work was begun from drawings passed to the Great Baddow Research facility by the Marconi Osram Valve Company, which were in turn copied from the German Stabilivolt. There followed a six weeks' struggle out of which came the first very small batch of Marconi-made Stabilivolts.

These were taken by car to London for testing but en route, one was broken, a heart-breaking mishap, considering the labour that had gone into the work. Under test, it was also found that only half the samples were good. But by October 1941, under the direction of an senior engineer with thirty girls, many difficulties had been ironed out, and the output ramped up to thirty units per week under an Admiralty contract.

But far larger numbers were required, and to cope with this demand, work was transferred from Great Baddow to the New Street factory in Chelmsford and the (EEV) factory in Waterhouse Lane . There, for the remainder of the War, and afterwards, Magnetron, Stabilivolt and valve production has been carried on.

Throughout the War, no bomber flew that did not carry Marconi equipment of one kind or another and few Spitfires or Hurricanes flew that were not equipped with the Company's Stabilivolts. When full production was achieved, the output of Stabilivolts was 8,000 a month and over 350,000 were manufactured during the war.

Throughout the war work continued day and night unceasingly. Very early in the War, the standard wireless 'Service' sets were either converted to, or replaced by, crystal-controlled sets. This immediately created a very large demand for crystals. It was not a question of thousands, but of millions of crystals, and so a new field of experiment, organisation and production was opened up.

Marconi's had been pioneers in the use of Piezo Electric Quartz for the frequency control of high-frequency oscillators. In 1926, a crystal for a frequency of about 94 kHz was cut by hand, a job that took a complete working week. Development and application were continuous and by 1931 a Marconi broadcast transmitter was fitted with a zero temperature co-efficient crystal which was the first of its kind in Europe. This station was in the Marconi Works and was operated by the Company for the B.B.C. Two years later, some twenty-two European broadcasting stations were fitted with crystals, and so real frequency stability was virtually achieved.

In 1942 a separate Department was set up to deal with the enormous increase in the demand for crystals. Its pre-war crystal manufacture had been carried out under laboratory conditions whose techniques were not suited for mass production. The new Department had a mushroom growth which at the peak of output employed over two hundred people, spread over two shifts, twenty hours a day, six days a week.

Early in 1940, a call came for additional apparatus to cope with the new activities of the B.B.C. and the Company received an urgent request for a new 2kW transmitter. There was no time to put the work through the usual channels of production and consequently a set was assembled in the laboratories of the Development Section from existing stocks of materials and experimental components originally intended for other purposes. The transmitter was eventually housed in a curious-looking wooden frame-work somewhat resembling a model cathedral, and soon after its delivery to the B.B.C. it was christened the Gothic Set, by which name it is still known.

During the War, more than half of all transmitting apparatus used by the B.B.C. came from the Marconi Company. These included transmitters operating in parallel, twenty in all, medium and short wave, erected at Droitwich, Start Point, Hull, Penrith, Redruth, Burghead and Lisnagarvey. Broadcast transmitters were supplied to the BBC from New Street to send out nightly messages into occupied Europe for the Resistance fighters and raise the morale of the oppressed civilians.

In 1942 the Royal Navy received a new centimetric radar system (Type 271) to which Marconi's made a significant contribution using the new Magnetron valves. Together with airborne radar, these sets played an important part in winning the Battle of the Atlantic, waged against the U-boats directed by Admiral Doenitz. Marconi's had already put in place the radar stations to provide the Air Defence Network a

The TR 1154/1155 transmitting-receiving set with which each British bomber and every Coastal Command machine that flew in the war was equipped, was a standard unit with many standard components. In October 1939, the R1155 receiver was merely 'something on the back of an envelope'. Designed at Writtle but first produced at New Street the first model was flight-tested by January 1940, and complete equipment was installed in aircraft by teams of the Company's engineers by June of that year. During that period of intensive production, ease and speed of manufacture became paramount considerations, and thus the ideal design was seen as that into which could be incorporated the maximum number of standard components

The growth of the importance of electronics over the forty years since the turn of the century is astonishing. In the Boer War wireless equipment had played an insignificant role, useful on occasions while in service with the Royal Navy, but its use being in no degree critical to the outcome. World War I had seen its status raised from that of a mere 'adjunct to visual and telegraphic signalling' to an indispensable means of communication and co-ordination for all the armed services. By 1940 wireless was the key technology in winning the war.

During the First World War Marconi had pioneered the development of direction finding (D/F). In WW2 considerable effort was expended on identifying secret transmitters in the United Kingdom by direction finding. The work was undertaken by the Radio Security Service (RSS also known as MI8). Initially three Marconi-Adcock HF D/F stations were set up in 1939 by the General Post Office but with the declaration of war, MI5 and RSS developed this into a larger network.

One of the problems with providing coverage of an area the size of the UK was installing sufficient D/F stations to cover the entire area to receive skywave signals reflected back from the ionised layers in the upper atmosphere. Even with the expanded network, some areas were not adequately covered, so up to 1,700 voluntary interceptors (mostly radio amateurs) were recruited to detect illicit transmissions by ground wave. In addition to the fixed stations RSS ran a fleet of mobile D/F vehicles around the UK. If a transmitter was identified by the fixed D/F stations or voluntary interceptors, the mobile units were sent to the area to home in on the source.

By 1941 only a couple of illicit transmitters had been identified in the UK. They were German agents that had been 'turned' and were transmitting under MI5 control. But many illicit transmissions had been logged emanating from German agents in occupied and neutral countries in Europe. The traffic became a valuable source of intelligence and the control of RSS was subsequently passed to MI6 who were responsible for secret intelligence originating from outside the UK. The direction finding and interception operation increased in volume and importance until 1945.

The HF Marconi-Adcock D/F stations consisted of four 10m vertical antennas surrounding a small wooden operator's hut containing a receiver and a radio-goniometer which was adjusted to obtain the bearing. Medium Frequency stations were also used which used four guyed 30m lattice tower antennas. In 1941 RSS began experimenting with 'spaced loop' direction finders developed by Marconi's and the UK National Physical Laboratories. By 1944 a mobile version of the spaced loop had been developed and was used by RSS in France following the D-Day invasion of Normandy. The US military used a shore based version of the Spaced Loop DF in WW2 called 'DAB'. The loops were placed at the ends of a beam, all of which was located inside a wooden hut with the electronics in a large cabinet with cathode ray tube display at the centre of the beam and everything being supported on a central axis. The beam was rotated manually by the operator.

By the late 1930s, the British had begun using high-frequency direction finding (HF/DF or 'Huff Duff') equipment on their warships. During the Second World War, the Royal Navy successfully used introduced a variation on the shore based HF/DF system to locate German submarines in the north Atlantic. They built groups of five D/F stations so that bearings from individual stations in the group could be combined and a mean taken. Four such groups were built in Britain at Ford End in Essex, Gonhavern in Cornwall and Anstruther and Bowermadden in the Scottish Highlands. Groups were also built in Iceland, Nova Scotia and Jamaica. At the end of WW2 a number of RSS D/F stations continued to operate into the cold war under the control of GCHQ.

The Second World War had seen the beginnings of an entirely new concept in which battles waged between opposing forces electronic impulses could bring victory or defeat. Perhaps the most spectacular instance of this was the Battle of Britain in which radar played so decisive a part, but throughout, the armed services electronic devices for communication and control purposes became indispensable units of the war machine.

The hundreds of receivers and antennas at wireless monitoring stations throughout the UK along with the 'voluntary interceptor' listeners fed continual information to the code breaking centre at Bletchley Park, whose crucial intellignec output undoubtedly turned the tide of the war. The 'battle of the beams', fought in secret between radio engineers and scientists on both sides defected thousands of tons of bombs away from their intended targets and saved countless lives. Winston Churchill wrote:

> 'During the human struggle between the British and the German Air Forces, between pilot and pilot, between AAA batteries and aircraft, between ruthless bombing and fortitude of the British people, another conflict was going on, step by step, month by month. This was a secret war, whose battles were lost or won unknown to the public, and only with difficulty comprehended, even now, to those outside the small scientific circles concerned. Unless British science had proven superior to German, and unless its strange, sinister resources had been brought to bear in the struggle for survival, we might well have been defeated, and defeated, destroyed.'

In any modern war the 'final issues' still have to be settled in terms of human lives committed to the conflict; men and women who have to fight in the front line.

But crucially, the conduct of any war relies on the people at home who make the equipment that has to go to war.

Memories - Marconi's during World War II

'**In** August 1941, I joined the Staff of Power Test at a yearly salary of £156 first working Marine Equipment including Lifeboat Spark Transmitters. At that time Test was treated as an extension of Development so I was able to help A.W. Lay with his Medical Diathermy Machines which because they allowed bloodless surgery were used to save lives of hundreds of soldiers wounded in the battle-fields North Africa. Later in the War these Spark Diathermy machines were used to jam the German Bomber navigational systems.

The labourer in Power Test was Fred Hutchinson who in 1922 had found a chair for Dame Nellie Melba to sit on for her historical first broadcast: Many of the established Test Engineers were characters. Sandy practiced economy by having his hair cut only twice a year, another was known as Fur Coat Baker after he had brought one back from an installation job in Switzerland and wore it every winter. Mr George insisted on having his desk in the high voltage enclosure, until some eight foot sparks flashed around him.

Several of the senior installation engineers like R.N. Vyvyan (one of Marconi's earliest engineers) had impressive profiles. It was said that an early Chief Engineer, Andrew Gray, would only employ engineers with large noses as he believed this indicated strength of character.

The main output was the SWB8/11 (Short Wave Beam) range of transmitters of which some 800 were produced to maintain War-time worldwide communications. After each transmitter was tested, it was put on a load-run for two hours. These periods were enlivened by tales of foreign parts by visiting installation engineers. One, for example, had just returned from the Amazon jungle where hundreds of miles from the nearest hospital; his colleague was struck down with prostate trouble. The local village Doctor decided to operate and taught the engineer how to give Chloroform on a pad. Before the Doctor had time to sew up the incision the engineer had to grab a bowl to catch the flow resulting from the relief of pressure on the patient's urethra. The last drop just filled the bowl held over the still open incision. The colleague made a good recovery.

The Franklin Drive in the SWB8 transmitter was a variable LC oscillator using a very clever mechanical design which allowed precise frequency setting and freedom from drift due to temperature changes. This used a DEQ (Dull Emitter Q) valve, designed by Captain Round in 1919 as a narrow band FM modulator to reduce the effect of selective fading.

When Marconi sent one of the SWB series to the USA for evaluation their engineers used one of their latest Quartz crystal test sets to check the transmitted frequency. When no audible beat was heard they laughed and said the Franklin must be no better than could be expected of an old fashioned LC oscillator. So while this test was under way the Marconi engineer slowly adjusted the Franklin out of zero beat.

Later in the War, the SWB8s were modified so they could be used to 'bend' the navigation beams used by the German air-force so their bombs fell harmlessly on moorland or the open sea. A special SWBX version was also produced to counter the guidance system used in the early Flying Bombs. To get these out in time the Test staff worked 24 hour shifts.

There was a story that two of these transmitters were used tuned approximately to the correct HF waveband but spaced by the IF frequency of a flying bomb used against allied ships in the Mediterranean. As a result an Italian battleship sank itself. One night Admiral Grant came into Power test shouting 'Cease Fire', they all thought the War was over, but it only meant that the Germans had changed their guidance system so that the SWBXs were no longer needed.

Some SWB8s were mounted in mobile vans, possibly for use at the Churchill, Roosevelt, Stalin conferences. These were tested up the Yard at night. As the job was urgent the Test Engineer decided to relieve himself over the tail-board of the van, unfortunately he had forgotten to earth the vehicle which was wired up to the 400 volt mains.'

Peter Helsdon.

'In the fields near the factory was a Barrage Balloon site plus airmen. We used to spend our dinner break in their hut drinking tea etc. often watching the Balloon go up and down if there was an Air Raid.

We had Ration Books
2 oz tea
4 oz sugar When available (I have drunk tea without sugar from that day.)
2 oz butter
4 oz meat
1 egg per week
1 box Powdered egg.

I never handled Ration Books. We never learnt to cook partly as there was no time and food could not be wasted. There were Barrage Balloons right up to the outskirts of London.

We had a big Anderson [probably *Morrison*] shelter in the house about 10 ft square, steel topped with mesh sides and we used to sleep in that if the sirens went. It had to serve as a table because it took up all the space. We often had evacuees down from the East End. Mothers and small children. They complained about being in the country, hated the fields and scuttled back to London as soon as possible. We also had Service men billeted in our houses - Polish Airmen - Soldiers etc. They came to Marconi's on courses. Radar was just being invented. They would stay for about a month and wives would come too if they were married. As they were Armed Forces they received extra rations.

Bulletins came over the factory radio and also we had Workers Playtime every day. I worked amongst about 100 girls and everyone sang to the latest songs with Vera Lynn, Ann Shelton and hosts of other stars.

This was the time of 'Doodle Bugs', aircraft with no pilots which crash landed on London and Southern England day and night. Although we felt we were winning the war these were unknown nightmares doing a lot of damage. '

Olive Cox

'I was born in Grays and worked for Singer Sewing Machines in Clarence Road but went to live with my sister in Chelmsford during the war; I was allocated to Marconi's in 1940. Cecil Myhill was my boss and Gerry Scriber was his assistant. I was put with Peter Brett to help with a small wireless set that would eventually go into a very large set.

For two years I was Hilda Sibley then in 1942 I became Hilda Rounce when I married George. Those days were amongst the happiest in my life. Everyone was kind and helpful because I knew nothing about wireless sets. If I did not know the name of a component that was needed I would draw it and take it to the store. I worked later with Nancy who later married Cyril Miles. Both she and Joyce Fairhurst were good friends. I remember Dora, Katie, Barbara and many more.

We worked next to the Tool Room where Danny would toast our cheese sandwiches in his machine. Happy, happy days. My sister Ruth was in Mr. Stock's office; a lovely man. Most of us cycled to work and at the end of the shift New Street would be full of bicycles; no cars in those days. I left in 1944.

When I visit my son in Chelmsford he sometimes drives me past the factory and it seems such a shame to see it boarded up and derelict. The double gates we used to cycle in and out of are still there and evoke so many happy memories. At 92, I would imagine I am among the few who are alive today.'

Hilda Rounce

Just before war broke out a record of the Marconi Wireless and Telegraph Company male juvenile rates of pay dated 29th December 1936 was published. A piecework bonus could also be available. A 25% additional amount could be paid in some circumstances when the piecework rate was not available to all.

The chart shows that workers as young as 14 started their careers at New Street and that by age 15 onwards two grades of staff were recognised with different pay rates, an additional ability rate and a food allowance. These are for a week of 50 hours. In 1936 the average pay for an agricultural labourer was around £1 12 shillings per 50 hour week and a skilled craftsman about £4 per week. [Before decimalization on 15th February 1971, there were twenty shillings per pound and the shilling was subdivided into twelve pennies.

Age	Grade	Basic Rate s	d	Ability Rate s	d	Food Bonus s	d	Total s	d
14	1	6				4	6	10	6
15	2	7		1	6	4	6	13	0
15	1	7				4	6	11	6
16	2	9		1	6	4	6	15	0
16	1	9				4	6	13	6
17	2	11	3	1	6	4	6	17	3
17	1	11	3			4	6	15	9
18	2	13	6	3		4	6	21	0
18	1	13	6			4	6	18	0
19	2	18		3		4	6	25	6
19	1	18				4	6	22	6
20	2	22	6	5		4	6	32	0
20	1	22	6			4	6	27	0

Hoffmann's, Crompton's and Marconi's all had their daily hooters, both at clocking on and clocking off times. Although the Companies staggered their working hours at mid-day and by 5.15pm New Street exploded into life as hundreds of cyclists rode over the cobbles and crossed the sunken railway lines leading from the goods yard directly into the Marconi works.

Through the narrow railway arch or up Rectory Lane men and women furiously pedalled home for a hasty lunch or their evening tea. Many buses also lined up in Rectory Lane and New Street to run workers home.

Cycles en masse

'I can remember that because in those days they recorded that Chelmsford, you know, there were more bicycles in Chelmsford than there were even in China today! But that is an exaggeration of course, but if you.... 12 O'clock when Hoffmann's and Marconi used to come out, all the way up past the police station and what have you, they used to come up there down into the High Street and you just could not move for bicycles. Everybody had cycles. Of course I forget how many employees Marconi and Hoffmann's had.... there was no Canteen to shop in, they had to cycle home at midday and of course 5 O'clock at night they used to cycle home again.'

C. Frost

'**My** memories, all those double-decker buses, the very narrow railway bridge, and hundreds of cyclists all wanted to get home.'

Mr P Betts

'I think one of the most frightening things in those days was usually on a Friday night and it usually rained on a Friday night and there were leaving the Company possibly a thousand cyclists, push cyclists ...or more than that because they were coming down from Hoffmann's as well and they were slightly cobbled stones on the street, the railway bridge that went over the road was very very narrow, the pavement was about 3 foot wide and err at the same time they were bringing cows down to the goods train from the market so you got about 30 or 40 steers running down the road in the opposite direction to the cycles and people flattening themselves against the railway bridge . . .

and if you want a definition of hell I think that was it. Used to scare the life out of me on a Friday night.'

David Newman

'**George** Wilson (1921–2004) commenced his four year apprenticeship in instrument making with Marconi's Wireless Telegraph, Co, Ltd on 10th September 1938 at the age of 17. His weekly wage for the first year was a basic rate of 11/3 plus food bonus of 4/6, totalling 15/9. His apprenticeship was completed to the Company's 'entire satisfaction' on 10th September 1942.

During this period he met his future wife Kathleen Shelley, who also worked in the Instrument shop at the Marconi Works, and they married in 1944. They were a popular couple and at least 80 colleagues contributed towards the £5.13s wedding collection

In the war years George also served with the Marconi Bomb Disposal Unit and between September 1941 and October 1944 they dug up 41 bombs. George enjoyed the annual works outings to places such as Hastings, Southsea etc. He also helped with the Marconi Provident Society which assisted contributing members in times of sickness and distress.

George took early retirement on 22/12/1983 and was presented with a wall clock by John Sullivan who used to be the Purchasing Manager of Marconi Communication Systems Limited at New Street.'

Susan Wilson

Marconi Bomb Disposal Group

Marconi Company Outing, Hastings, June 1955

'**My** father was Cyril Edmund Jackson, and he worked for 40 years in the Drawing Office at New Street, the last part as a Section Leader. As well as being a senior Air Raid Warden, he became a *'Spotter'* on the roof at New Street all through the war, wearing a Home Guard uniform and learning to fire a gun. He was on the roof when a daylight raider dropped bombs around the Hoffmann/Henry Road area.'

Eunice M. Jackson

Because of its essential, war-related industries and the importance of its railway, Chelmsford received many bombing raids during the Second World War. Along with the Marconi and Hoffmann's factories, which were in the town centre, the Crompton Arc Works and the adjacent rail viaduct were conspicuous and strategic targets.

In an attempt to disguise the Marconi factory from the air, the building was painted in elaborate camouflage to both break up its aerial pattern and try and make it resemble fields. This included painting the roadway in the sidings blue to look like a river and the water tower was painted to look like a church steeple complete with 'stained glass windows'. The camouflage and church window paint is still visible even today.

Joan Wigley joined Marconi's when she left school at 14½ and remained with the Company until she took early retirement in 1981. In her time she was responsible for staff records, apprentice and staff training administration at New Street, Marconi College, Baddow and EEV Waterhouse Lane. From 1942 to 1943, under compulsory call-up to industry for females aged 20 she carried out testing of components in Test Department. The following extract is her memory of serving in Marconi's ARP Gas Decontamination Section during the war years.

I heard the loud bang!

'I had to learn all the gasses and treatment needed if gas was used in warfare. We rehearsed in protective clothing and Service gasmasks and had to test them in a van of gasses! On call duties each time the air raid siren went during working hours, off I went up the yard to the underground shelter with my respirator and tin hat to await the 'all clear' then back to work. I was on call one evening a week 6pm – 12.

Many a time I cycled through Chelmsford during a raid with tin hat on and respirator on my handlebars! Marconi House had some bombing and one day the windows were broken and several of the staff on various floors formed a human chain and passed buckets and baskets of broken glass etc down from top floor to ground! Little did we know at the time that an unexploded bomb was just a few yards below us! We were sent home at once. A few hours later it had been detonated. At home I heard the loud bang!'

Joan Wigley

The New Street Works had several air raid shelters, their exact location now lost. During the First World War the Company dug trenches to protect against Zeppelin attacks approximately where the bicycle sheds by Building 720 were located, hoping that the railway embankment would provide additional cover. One Second World War shelter was discovered when Eastwood House construction began, located roughly below Section 16 in the shadow of Marconi House, another was 'adjacent' to Building 720.

'I worked in an office in a dairy in Baddow Road but when we were 18 all youngsters were called up and interviewed to see if they would go into the Forces or into the factories.

I had younger sisters at home who needed whatever I could earn as both my parents and elder sister had been killed earlier in the war. I went to work at Marconi's, Chelmsford in the Valve Laboratory. My job was joining the metal to the glass. They both had to be heated to join together. Rather like making pastry was how it seemed to me at the time. People were very kind and patient showing us all the work. Because people were called up they came from all over Essex and Suffolk and lived in digs.'

Vera Crause

'I lived in Chelmsford with my mother and sister because my father was in the 8th army. He served the 6 years without injury and came home about twice in that time, so my mother had full responsibility of us girls. Because Marconi's was situated about 100 yards from where we lived, it was prime target for bombing. Every night when the sirens went off we were not allowed to stay in our home.

So therefore, the ARP wardens would escort us to the Shire hall in Chelmsford where we would have to stay until the all clear went. Unfortunately Marconi did get a direct hit and the Germans seem to leave us alone after that for a short time.'

Mrs Summers

'I would leave my house in Wickford at 4 p.m. and cycle to Chelmsford in the dark. I worked at Hoffmann's. My job was to view the bearings for Rolls Royce. This involved checking them to make sure they were of a certain standard and not rusty. If they were no good they were rejected. I worked at Hoffmann's until part of it was bombed and then I got a job at Marconi. At Marconi my job was to wind coils. We were not sure what they were used for. I always worked nights.

I had a daughter, who was nine. We lived with some friends and my daughter was looked after during the day. When I think about the war, I think about Chelmsford being really bombed and it was bombed so badly that we all had to walk home from Chelmsford because the town was so badly destroyed.'

Joan

Early in the morning of 9th May 1941 a single German aircraft carried out a raid on the Chelmsford New Street Works, scoring direct hits with three of the stick of four bombs dropped. Seventeen employees were killed and forty injured. The houses at numbers 70, 72 and 74 in Marconi Road were also destroyed with another two people killed. The first bomb detonated in the middle of the main Machine Shop; the second hit the Transmitter Erection Shop, wrecking it and an adjacent electricity sub-station and setting fire to the Paint Shop. The Company fire brigade quickly got the fire under control however, while a labour force hastily improvised a new sub-station from spare equipment stored in another part of the Works. But on the following day a ticking noise was noticed coming from under the debris of the transmitter erection shop. It came from a delayed action bomb which had to be blown up in situ and in so doing the temporary sub-station was completely destroyed.

Production was badly affected for a short time but temporary quarters were soon found, including a large Carpenter's Shop which was lent by Crompton Parkinson Ltd., of Chelmsford, pending the erection of new buildings at New Street. After a comparatively brief interval, production climbed back to 80 per cent of normal output.

The Bombing

'It was the machine shop, I think, twenty odd people killed, and I can remember walking past.... and didn't know but there was another bomb there, anyway I went back to the Test Department where I was working and I was down busy and one of the air raid people came round and said 'What are you doing here?' Apparently they had found another bomb and I was the only one not in the air raid shelter! So I scuttled down and a few minutes later the bomb went off, because they couldn't move it so they... and that doubled the amount of damage, but of course no one was hurt the second time.'

Geoffrey Nash

'I moved to Chelmsford, two weeks after the start of the War from my home at Reigate, Surrey at the age of 24 having lost my parents that year and to be near my sister whose husband had a shoe repairing business opposite the Two Brewers Public House in Springfield Road. I applied for a job in Marconi's as a Skilled Centre Lathe Turner and was engaged in a reserve occupation. Due to the work load a Night shift was introduced.

I met my wife at the end of February. She was Maternity Night Sister in Charge at Wood St. Hospital St. Johns, and we subsequently got engaged to be married in June, but this was not to be. I elected to stay permanently on Night duty as my wife to be was also on nights.

On the 9th of May we were having our lunch in the Girls Pavilion which was a wooden building where Building 720 now stands.

The Air Raid warning went and was ordered by the Home Guard to leave, however instead of going to the underground shelters adjacent to the building we went into the Works to a place at the end of Dept HW1, which is now the Air or Raw and Wire Stores. We were split into groups in various parts of the factory. We all had our helmets and gas masks on and at the dispersal point was a dart board on which several men played darts.

I remember hearing the bomb coming and threw myself down alongside of a truck which was used to move lengths of rod from the Raw Stores to the Capstan Section. I came too some time later in one of the underground shelters with Mr Alf Young who was the works first aid man although it was dark I knew his voice.

He said 'All Right Alf the doctor's coming'. I said tell him to B@-well hurry up. I am shaking with cold as I thought but this was due to shock. Apparently I was moved to London Road Hospital unconscious at about 5.30am. I came to at about 9.00am and a visit by my future wire who phoned up the Works to enquire if I was a casualty. I said what are you doing here at this time of the morning. She said she had come to see how I was.

I suffered 3rd degree burns to my buttock and arm. Also discharge from my right ear. I was told that the bomb a 500lb had dropped in the Paint Shop, just the other side of the wall from where we were and the building caught fire.

On Sunday Admiral Grant and management visited the casualties in the Hospital. Monday morning we were told that a further unexploded bomb had been found and that if we heard a bang it was due to being blown up by the bomb disposal unit. I felt the bed jump up but did not worry as we knew what the bang was. I spent 5 months in hospital and was treated by Mr Harris. I had two skin grafts and eventually was sent to Bellfields which was adjacent to the hospital. During this time in hospital we were paid our wages at the piecework rate less the casualty allowances of about 22/6 per week which was subsequently refunded.

My wife and I subsequently were married at St. John's Church on December the 6th – the same day that Pearl Harbour was bombed. Due to the care of nursing staff my wounds eventually healed but I had lost the hearing of my right ear with 98% perforation. I received a pension of 7/6 a week which was eventually to 13/6 per week as married man. In 1946 I attended an examination by medical officials at the rear of the Saracen Head P.H. and was given £45 to write off the pension. They did not consider I was suffering any disability. What price for the permanent loss of hearing for the past 40 years!

Before restarting work I was seen by the Work's Doctor – Dr. Kerr and on referring to my ear he said I do not know anything about ears 'start work on Monday'. As there was no machine available on days so I had to start work on nights again and worked through the rest of the War. A lot of the time 7 nights a week – week in week out. Although my wife was a midwife we lost our first son at 4 days and her subsequent confinement in November 1944. I walked from St. John's hospital as there was no bus and arrived at work at 8.15am and was reprimanded for not being at work at 7.45!! '

A.J. Brown

'...**that** was on 8 May 1941. That was when the Germans came over and they ran a stick of bombs and they bombed Hoffmann's as well... I think there were seven or eight burned to death in Marconi's, mostly girls from the winding shop because they got behind the glass wall and then the fire bombs came down and everything was alight and you could hear them screaming and they couldn't get them out, the rescue men and that was a horrendous night because everything was wrecked and I mean wrecked and Marconi Road alongside so many of those houses were absolutely demolished and little children running around in the road in their little night clothes and bare feet and getting their feet cut to ribbons on broken glass and all sorts of people buried under the rubble, it was horrendous.

We were in the house 10 weeks and my husband was called up. I let the house and went to live with my mum and dad and got work at Marconi's, working on a very large machine, a Herbert 4 lathe. It was very interesting, putting a bar of material (steel, brass among others) working all these different tools to make all these different things. One was very special and I had to keep the blueprint hidden, these were making the start of the anode for radar.

I worked there all through the war and was on night shift when we were bombed, losing 17 of my work mates – all men. Even though the war was on we used to love working night work as we had music going on all the time. We took our special records in and the man would play them. The girls used to wear coloured berets to match the blouses they wore under their dungarees – we certainly kept our spirits up with all the singing and the few girls in their different colours.'

Alice M Atkinson

'A friend and I got a job at Marconi's on the machines. We worked on Christmas Day as our men were away. During a raid the warning had gone and we went behind the blast wall built just around the machine shop. We were all lying on the straw palliasses that the soldiers had given us, they were manning the guns on the roof. We were singing the war songs as usual when along came the Night Manager and he went berserk, there was a man and a woman lying on the same mattress cuddling. 'Right that's it, tomorrow girls over the other side of the machine shop and the boys stay here'.

Less than a week later the bombs fell on the shop and 17 of the lads were killed. The next day we went to see if we could go to work that night. No we couldn't, so three of us girls got a job loading some of the bomb debris on to a lorry. When it was loaded us girls got up on the lorry and were taken to the site in Waterhouse Lane and helped to lay out the rubble which formed part of the foundations of the E.E.V. which is now Marconi's.'

But the raid had a very human cost, counted to this day. Joan Griggs was 23 years old when her husband was among those killed. She had been married for less than two years to Alfred Griggs, 28, an engineer and one of the youngest of the 17 workers who died. Some 68 years on, Joan, from Writtle, was among the proud relatives attending a special ceremony at Chelmsford Cathedral to unveil a plaque to remember the Marconi night shift employees who lost their lives in the bombing, which caused a huge fire at the factory.

'**People** say it was a long time ago, but you still have the same heart,'

'I couldn't believe it when I found out Alfred had died. We had only been married for one year and 10 months. We had just bought a house and were hoping to start a family. It was very difficult for me and for years I had nightmares about fires.'

'I am pleased about the plaque, but I wish it had been done a lot earlier. I would have liked my contemporaries to see it, but I have lost them all now.'

Joan Cockerton

Several other families who lost loved ones in the bombing of the New Street Works factory were at the ceremony, including Dawn Swindells a legal secretary from Washington DC, whose maternal grandfather Charles 'Tom' Franklin, from Coggeshall, was killed aged 46. Her uncle Victor Joslin also died, aged 31, but her father Bert Joslin escaped. Dawn gave a reading from the New Testament at the ceremony and among those in the congregation were her mother Joan Joslin, now aged 87, from Coggeshall, with other family members.

'**My** father was such a loveable man and worked so hard as a Sprayer in the Machine Shop. I was 18 at the time and just burst into tears when I found out he had died. It does give you a chill to think of that night, all that time ago.'

Dawn Swindells

The plaque was originally in the Marconi New Street factory and was retrieved by the Marconi Veteran's Association before the factory closed. It was remodelled thanks to an anonymous donor and it was decided it should be installed in Chelmsford Cathedral, below a plaque which honours victims of air raids on the Hoffmann's factory.

> **'The** plaque remembers the price Marconi personnel paid in the loss of their lives. They will not be forgotten, we will always remember them.'
>
> 'Marconi gave so much to Chelmsford and the world in the field of communications.'
>
> *Peter Turall MBE*

The plaque was unveiled by Lady Telford, widow of the Life President of the Marconi Company, Sir Robert Telford. It was formally dedicated by the Dean of Chelmsford Cathedral, the Very Rev Peter Judd, who in his sermon called for Chelmsford's statue of Guglielmo Marconi to be placed in the middle of the town, instead of its present location in a new square near the Civic Theatre. The plaque unveiling coincided with the 135th anniversary of the birth of Marconi.

The tragic bombing of the New Street Works on 9th May 1941 was only part of the story of Marconi's and Chelmsford at war. In 1940, on 13th October, a high explosive bomb fell on the house that stood on what is now the site of Christchurch, killing the Mayor of Chelmsford and his family. To the South, on 11th March 1941, another high explosive bomb had scored a direct hit on the approach road to the railway bridge in Princes Road. On the night of April 14th 1943 a wave of twenty bombers came in over Clacton and followed the railway line to Chelmsford. Here they dropped flares and carried out dive bombing using mainly incendiaries. Woolworths shop was burnt down.

The following month, on the night of 13th May, thirty enemy planes made a low level attack on Chelmsford. All the town's defences went into action. Twelve Eastern National Buses at their Depot in New Writtle Street were set alight. The Radar on the Cricket Ground was put out of action by bomb splinters but the Battery continued to fire under visual control. The Territorial Drill Hall in Market

Road was set alight by incendiaries and the ammunition stored there exploded all through the night throwing burning debris over the Battery site. One bomb landed on sand-bags protecting the Lewis Gun at the site entrance, but failed to explode. It fell at the feet of two Regulars and a H.G. man. Two Home Guard men were awarded Certificates for their courageous actions that night. Other bombs fell round the site, some near the path by the river. Houses in Park Rd. near the 'Z' Battery H.Q. were destroyed. All the Radar men were wounded but there were no serious Home Guard injuries.

About sixty people in Chelmsford were killed and many more injured that night. In Chelmsford there were many fires, and witnesses saw the prison and adjacent houses burning. There were bomb craters in a line from the S.E. corner of the battery site and across the river towards the radar cabins. Hawke's sweet shop on the corner of Victoria Road and Duke Street and Cannons restaurant opposite were destroyed. The Suet factory in New Street caught fire and the melted fat ran across the road, making it difficult to get to work at Marconi's factory next morning. None of the main factories were hit.

> 'I worked at Hoffmann's and Marconi's during the war years. I lived close to Hoffmann's, so we were affected by the bombing. We lived opposite Archer's suet factory in New Street, next to the Cathedral school.
>
> We were bombed one night, when fire bombs (incendiaries) hit Archer's. That went up in flames and the fire engines dealt with that. We had one in our bedroom that had come through the roof onto the bed. Unfortunately my mother's family bible was destroyed. We ran upstairs and threw the burning mattress out through the window. We had just managed to put it out when the door was thrown open, firemen rushed in with a hose and they were covered in suet from the factory. The firemen did more damage than the fire. However, we lived to tell the tale.'
>
> *Eileen Hance*

On 14th May 1943, six high explosive bombs fell across an area bounded by the railway and New London Road, from Hayes Close south westwards to Cherry Tree Lane. On the same raid two parachute mines fell flattening the Marconi

Pottery Lane factory in Broomfield which consisted then of nine various sized single storey buildings sited in the centre of a 10 acre field accessed from Pottery Lane. Previously a testing station, it had just been adapted to be a dispersed New Street site for making CR100 anti-submarine devices and receivers with 70% to the Admiralty. Normal complement was 150 staff, producing 100 sets a week at £48 10s each. Work had to return to New Street and by the end of May the CR100 work went to Waterhouse Lane and then to Bettawear Brush factory in Romford from mid-August 1943.

> '**Cyril** switched off the light, bent low and clambered into the shelter and fell asleep. He suddenly awoke feeling uneasy, went outside the shelter and heard a quiet singing sound and realised it was the sound of the barrage balloons cables being pulled taut. He knew that the balloons had been sent up high and that meant trouble. He and his mother quickly dressed and looked out the back door joined by their neighbour Mrs Hart and her son Ken. About 2 miles to the south west a brilliant light filled the sky and realised this flare was coming their way. Having gone back inside and taken shelter they heard an aircraft engine and at the same time a loud whooshing sound. Vibrating explosions shook the house followed by whistling shrieks, the whoosh of rockets and the crack of shells. Would it never end? And then — silence.
>
> There was a noise at the back door and there was Cyril's father in his uniform and steel helmet who had been sheltering in the Warden's Post checking to make sure his family was safe before going on duty. Cyril and his father went to look outside; all was quiet until they heard the sound of a plane flying fast and low. It grew louder and they felt the shock of its bombs falling even though they were dropped on the other side of town before it grew quiet again.
>
> The family waited anxiously for the All Clear then heard a crack and the lights went out. They saw flashes from behind the blackout followed by a heavy rolling rumble. It was a storm, it seemed that the Devil was having a last dance and we saw flashes of forked lightning around the barrage balloons setting fire to them one by one then falling to the ground like giant red Catherine Wheels.

At around 3.30am the All Clear finally sounded and when Cyril went to school later that day all his classmates were swapping tales of gore and bravery. Their headmaster gave them a stern lecture about discipline and air raids were not to distract them from their studies. When Cyril arrived home his father was there and his infectious spirit dispelled any fears that Cyril had, even more so when he showed him a 3 foot piece of shrapnel that was buried in the back door where they had been standing 10 hours before!'

Stephen Williams

'**We** lived in a bungalow in Broomfield, Chelmsford, Essex. My bedroom was used as the shelter room. Extra timber was put into the rafters to give extra strength above. At the window was a bookcase filled with books so the room was very dark. Also in my room was a Morrison shelter like a giant steel reinforced table. It was bigger than a double bed and lined with steel mesh sides. It was in a corner of the room and had a mattress inside.

In my room there was also a single bed for me. Over the door was a thick curtain to stop any intruding gas!! There was also a spare mattress under my bed. When the sirens went my parents shouted to me so I rolled under the bed onto the mattress beneath and went back to sleep.

If it was very noisy with guns and bombs my parents would come into my room and crawl into the Morrison shelter and pull over the wire mesh. Mum was very upset because our Labrador dog wasn't allowed into the Morrison shelter although she came into the room. When we heard a stick of bombs from the South East we held our breath. Boom, boom, boom, BOOM! Louder and louder. Sometimes they dropped on the nearby railway line. Chelmsford was a fair target.

I was 17 in January 1945 when a rocket dropped diagonally on Patching Hall Lane. There was a huge bang. I worked in the office at Marconi House in New Street, four storeys high on the top floor so I could see Broomfield and a tall column of black smoke rising from our area. I didn't know what to do so I ran down to my Dad in the instrument shop on the bottom floor. He couldn't vacate the building quickly as he needed a gate pass so he told me to cycle home to check if mum was O.K.

Part of me was terrified as I cycled home but I was also very curious to see what had happened. When I arrived Mum was sitting in the kitchen having a cup of tea with a friend. 'What are you doing, home from work?'

Oh, she was cross with my dad for sending me on my own. Fortunately there was no immediate damage, just a few damaged tiles. Nalla Gardens was the nearest hit to us. In December 1944, Hoffmann's factory (Rectory Lane) which made ball bearings was bombed killing about 30 night workers, mainly girls. There is a special memorial for them in the cemetery. Marconi's, Hoffmann's, the railway line and the river were all potential targets near to each other.

Barrage balloons were anchored to the ground to prevent enemy aircraft getting too low. When they broke loose they made a terrible racket. Sometimes the cable trailed knocking off roof slates and pulling down telegraph wires. They rolled around doing acrobatics in the air.

One landed four doors away outside our neighbour's French window. It seemed as if it was trying to break into the house. They were huge things. Sometimes they were struck by lightning during a storm and caught fire so when a storm was imminent it was a race to get them down. Really they were more bother than they were worth.'

Betty Morris

'I used to live in number 80 Marconi Road also and they used to have a balloon protecting Marconi's and Hoffmann's factory. But every time the balloon came down it took my Mothers chimneys down with it, she wasn't too happy.'

Hector Norman Smith

'**Hitler** must have heard of my birth; he chose that day to march into Poland. My first memories of the War, in fact my first memories of anything at all, were of the night that our house in Chelmsford was bombed. It was, I believe, in September 1942 when the bombers aiming for Hoffmann's Ball Bearing Works and the Marconi factory dropped a stick of bombs one of which landed on the back doorstep of our house in King Edward Avenue. I recall father telling me that Marconi and Hoffmann's were important places, but I didn't understand why.

When the air raid siren went, I can recall being bundled into the Morrison shelter in the downstairs front room of the house with my Mother. There was a big bang and I can remember the sounds and feelings of things falling about us. We were pulled out of the rubble by my Father and someone else. My Father had been at the top of the stairs when the bomb hit and was blown down to land at the front door. On the way, his head hit the arm of a Monk Seat at the foot of the stairs and snapped off the inch thick oak. The damage to the seat can be seen to this day. Father's head survived, relatively undamaged.

I can remember being passed out through the remains of the stained glass window of our front door into the arms of someone with a tin hat and in a dark uniform. There was a body lying in the street and I was later told that the lady who lived next door with her family had been blown out of the front of her house and was killed.

We moved to an Uncles' farm at Blackmore for some months until we rented a cottage at Oxney Green, near Writtle. We lived there for about six years and during that time I could always hear the air raid sirens going off in Chelmsford (about 3 miles away) before my Father did. He was a Special Constable, having served in the Devonshire Regiment and the Medical Corps in World War 1. I remember often seeing the V1s, or 'Doodlebugs' flying over us and, on one occasion, running in to tell my mother that the motor had stopped. She did not appreciate my enthusiasm! I remember seeing, but not hearing much of, one or two V2s as they went overhead.

During the winter months, school milk was warmed around the huge fire that stood in the classroom at Writtle School (I've hated warm milk ever since) and I recall the authoritarian Headmaster George Whitehead and his expertise with an old leather slipper. I remember my first orange, it was in a Christmas stocking one year and I didn't know what to do with it.

My Father came home one day tickled pink that he'd been able to get hold of half a parachute. My Mother made him some shirts and a blouse for herself out of the white silk. Mother was always making clothes for us out of all sorts of things. She once made me a brown beret which I hated. I put it in my pocket as soon as I was out of sight. She would have been very upset had she known.
If I was still hungry after a meal Father said I could have a slice of bread. If I was then still hungry I could have another slice, this time with a scrape of margarine on it. Only if I was still hungry could I have another slice with jam - if there was any to be had. To this day, I still clear my plate as I was taught to during the War. I rarely leave food.

My Grandparents lived in Colchester and I can remember visiting them in our old Austin 7 car and seeing all the Army vehicles and tanks parked down the middle of the A12 Arterial Road. I didn't see any enemy aircraft wrecks but some of my schoolmates used to carry bits around with them for weeks.

My Father 'captured' a German parachutist one foggy night - it turned out to be a tarpaulin over a haystack! Father had a African Grey parrot which survived the bombing of our house, and was found in its squashed cage 4 feet under the rubble and called out 'Hello boy!' to my Father who found him, but it died a few years later.

We moved back to a rebuilt house in King Edward Avenue in 1948, the year my Father bought one of the first Morris Minors. We had crowds of people stood out in the street looking at this new fangled motor.'

Ted Bocking

'I was born in 1928 and in 1937 moved to Chelmsford where my father was a police constable. My first war memory was my father being recalled from our holiday because of the Munich crisis. He started to dig a shelter in the back garden with some help from a young probationary policeman. As he swung the pickaxe he put it through my mother's fresh washing hanging on the line. After Chamberlain returned with his piece of paper my father filled in the hole he had dug!

My parents had arranged for me to go to Canada but when it came to the day of departure they couldn't bear to take me to the station so I stayed at home. Large numbers of evacuee children were billeted in our street and I remember being shocked at their behaviour, climbing lamp posts and smashing bulbs and other damage. They also had an unfortunate effect on my young brother who started speaking cockney and using swear words!

Although the police service was a reserved occupation most young officers volunteered. Only a handful of older experienced officers remained including my father — they had to lead and guide the wartime reserve policemen who were older men directed into the service. The War brought masses of extra regulations in addition to normal criminal law and also the police controlled the air raid warning and report service.

We became used to the sound of enemy aircraft and one night an aircraft was shot down over Chelmsford, my father brought home a large piece of metal which smelt like rotten cheese. My mother threw it out! The aircraft was found totally destroyed in the Bishop of Chelmsford's garden with the crew dead or dying. The next day inspecting officers from the RAF were most perturbed to find a lot of the aircraft lost to souvenir hunters which all had to be retrieved including my father's prize!

Because my father worked long hours I occasionally took food into him and once he let me see an German airman who looked very smart and spoke impeccable English. He told my father that as he was an officer he should have better quarters!

We saw the Blitz on London from our garden — the search lights and the sound of gunfire. When the oil refineries were bombed some 20 miles away we could see the fires. Although we did not feel in direct danger we spent most nights in the now re-dug shelter in the garden until winter came and it became cold and damp then my father built an indoor one. With tightly fitted blackout shutters and a coal fire it was quite cosy.

As boys we were obsessed with aircraft recognition and anything military and did not understand the politics of war but it was a different matter for our parents who had to cope with food and fuel shortages but there was a tremendous cooperative feeling with people working for the national good.

Women were being directed into war work but my mother with 2 young sons and a husband to look after baulked at 12 hour shifts but found herself required to take in lodgers instead. Every fortnight 2 or 3 airmen were billeted on us while they attended a radio course at Marconi's. She received 17/6d for each and 21/6d for officers per week. We must have seen over 100 of these airmen, we became particularly friendly with 3 Polish men who even spent their leave with us. I still have the plaque they presented to us before they left for France in 1944. In 1998 I tried to trace them and had some success with one who had settled in France after the War but had unfortunately died just after I started my search. His widow sent me a photograph of him holding my young brother in his arms taken outside our house.

After the War I was called up for National Service and found myself on the same radio course as those airmen who stayed with us. I was one of the lucky ones who was just too young to be involved in the conflict.

Stephen Williams

'I was at Moulsham Junior Girls' School from 1939-42. In the winter of 1937, we moved into a new rented house in Moulsham Drive. It was a bitterly cold day. There was snow on the ground and the road had not been made up.

My school life began at the beautiful new Moulsham School. I remember very little of the Infants' School – just small chairs and desks – but I loved being at Moulsham Junior Girls' School. I remember Miss Rankin as a fair Headmistress.

The war did not stop us having fun, but I did get annoyed when the siren went off during drama class, which I loved, but never during maths! The air-raid shelters were brick built, and had a peculiar damp, musty smell.

One air raid sticks in my mind, the one where Marconi's received a direct hit. We could see the flames from our back bedroom window. My brother and I stumbled downstairs, half asleep, to the newly acquired table shelter, which took the place of our dining room table. The shelter was made of steel, and very ugly. Once inside, with the wire sides pulled up, you felt like a rabbit.

Mum was good at cooking. She used to make what she called 'frippets', with dried egg. She would add water to the egg powder and mix until it was fairly thick, then dip slices of dry bread into the mixture and fry them. Delicious. The word 'cholesterol' was never mentioned in those days!'

Diane Berthelot (nee Lawson)

'I lived in Chelmsford Essex. I had five brothers, four were in the army and the fifth, who was medically unfit, worked at the Marconi factory all through the war.

After the Marconi New Street factory had taken two direct hits on May 8th 1941 they were clearing up the mess when he was handed an object and told to take it up to Galleywood Racecourse to the old main stand. The object turned out to be a severed leg of one of the bomb victims. Bodies were all taken to the racecourse.

I worked as a volunteer in the Y.M.C.A. Canteen, Victoria Road, Chelmsford. It was used by many service men. I stayed nearby with my sister-in-law in Highfield Road. One night I heard dreadful bombing and I was worried about my parents who lived near Marconi's and Hoffmann's. They weren't on the 'phone so I wanted to make sure they were all right. As I reached the railway station we weren't allowed through as part of the station had been hit so I went by way of Broomfield Road. I took off my high heeled shoes and walked barefoot.

By Marconi's near Glebe Road the area had been badly hit. Fire everywhere and ruined buildings, ambulances, firemen and rescue personnel. Eventually I reached my parents home where I found my family safe and unharmed in the shelter.

It was an exciting time for people of my age in their late teens and early twenties. We knew people were being killed and that was very sad.

It was hard with rationing and being bombed out but we felt we were really living. The pressure we were under was so high that we felt we had lived every minute of the day.'

Eileen Hance

On 19th July 1942, German bombs hit the Hoffmann works (which extended to the Railway Yard opposite the Marconi factory) killing 4 people. But the worst single loss of life took place on Tuesday 19th December 1944, when the 367th Vergeltungswaffe 2, (retaliation weapon 2) or V2 rocket to hit England fell on a residential area, including Henry Road and Rectory Road near the Hoffmann's Ball Bearing Factory and not far from the Marconi New Street Works which may have been the target. The New Street Works were protected by barrage balloons, but the V2 terror rocket was something new and terrifying.

From 12th June 1944 to the end of March 1945 Britain had come under heavy air attack from Germany's secret weapons – the V1 and the V2. The V1 flying bomb, known as the 'doodlebug or 'buzz bomb', was a pilotless aircraft with a one-ton warhead. By the end of June 1944, some fifty V1s were hitting London every day. Nearly two-fifths of the V1s were brought down before they reached their target, but over 3,500 got through, killing 6,184 people and injuring 42,146. The frequency of alerts and the ominous sound of the bombs cruising overhead made the V1 offensive an extremely trying ordeal.

But against Hitler's second reprisal weapon, the Vergeltungswaffe 2 or V2 rocket there was no defence. The first successful operational V2 was launched from Holland and landed in Paris near the Porte d'Italie on 8th September 1944. The same night, two V2 rockets fired from the Ardennes landed on the outskirts of London, one of them killing three people and injuring ten others.

On 12th October 1944, Hitler ordered that the V2 campaign should be concentrated on Britain and Belgium. Over 1,000 reached targets in Britain and some 1,800 were launched against Antwerp and other targets in Belgium. At its peak in December 1944, over a hundred V2 rockets a week were landing on the port of Antwerp. Unlike its predecessor, the V1, the V2 Rocket arrived unseen and unheard, delivering nearly a ton of high explosive at a speed of 3,500 feet per second.

More than a thousand V2s, each with a 1,000kg warhead, landed in Britain between 8th September 1944 and 27th March 1945. The attacks resulted in the death of an estimated 2,754 civilians killed with another 6,523 injured.

On 19th December 1944, the V2 rocket that fell on Hoffmans works killed thirty-nine people and injured 138, 47 seriously. From an Air Raid Wardens report:

> 'A V2 fell at 1.30am on Hoffmann's and Henry Rd, 40 people killed (29 worked at Hoffmann's) 115 injured, 1 large shop wrecked, 1 seriously damaged, general damage to Hoffmann's, minor damage to Marconi's, severe damage to Rectory Rd and Henry Rd 5 houses demolished, 25 seriously damaged, 403 minor damage.'

Bertie Upson was working at Hoffmann's Ball Bearings Factory when the German rocket hit. He normally worked 12-hour night shifts seven days a week throughout the war as a machine setter. He was on duty when a German V2 rocket annihilated parts of the factory, several dwellings in Henry Road were completely destroyed and many houses in nearby streets were badly damaged. Mr Upson and his team, about 80 women and a handful of men had just had their half-hour meal break when the rocket struck shortly after 1.30am.

> 'I was in the middle of setting up a machine and woomph! Things flew everywhere and the place was in total darkness. A few minutes later came the flames. I shouted at the girls to get out. Someone said the canteen hadn't been hit so I told them to go there. I kept four fellows back with me and said we had to do whatever we could to help.
>
> We got bodies and people who had been injured and half dragged and half carried them out to the Surgery. Sister Ham was in charge that night and, after taking several bodies down, she screamed 'don't bring any more, only bring people who are injured, not missing arms and legs', so we had to leave the bodies of the dead at the surgery door.'
>
> *Bertie Upson*

Mr Upson and his colleagues spent the rest of the night rescuing people from the burning rubble of the factory and several homes in Henry Road which had been 'smashed to pieces'. In total 39 people were killed that night. The soles of Mr Upson's shoes burnt through on the smouldering debris as he ferried the injured, dead and dying. Hoffmann's later gave him 30 shillings and seven coupons to buy himself a new pair.

> '**We** kept going until seven o'clock in the morning when American airmen arrived from Boreham in their trucks. We did quite a lot during those hours. We did what we could.'
>
> I went off and didn't go home for a day or two, where I was I don't know, but I just couldn't take it. A lot of them that were killed were young lads and such lovely girls.'
>
> *Bertie Upson*

The attack deeply affected him at the time and he still thinks about the many young lives that were lost that night.

> **'On** 21st December 1944 a two ton rocket dropped on the Hoffmann's War factory killing about forty people. My Father was then on nightwork and I went down with Leslie to try to find him. Most of the houses around were down or damaged. Christmas decorations were all over the roads. Lots of people trying to stop sightseers but knowing the layout of the different floors in the factory I left Leslie and eventually found my Dad shocked but unhurt. The nurses around said I could walk him slowly home which was about a mile away and we eventually arrived there. He insisted on going to work the next night but he seemed to have changed overnight.'
>
> **Olive Cox**

'When I was called up I was going to join one of the forces but changed my mind and went to work at Marconi in Chelmsford in the paint shop spraying wireless parts for the forces. For a time I lodged in Braintree with a girl I worked with. I then went to a lodging house in Marconi Road Chelmsford. I used to go home at weekends.

Nights out in Chelmsford were always fun. There were hundreds of American soldiers and air - men stationed close by and they used to swarm into the town. They always had lots of money, (not like our poor fellows). They would come to the dances, mostly at the Corn Exchange and teach the girls to jitterbug.

My sister Phyl, who was 2 years older than me was called up too she went to work at Hoffman's ball bearing factory., she worked nights and then travelled home, she preferred to do that. One terrible night on 19 December 1944 a V2 rocket was dropped on Hoffmann's, the very department where my sister was working and a fierce fire broke out. Thirty people including my sister were killed.

I was on nights at Marconi and as the two factories were opposite each other the noise and vibration were horrendous, I was thrown from my stool and somebody shouted, 'Hoffmann's has got a direct hit'. This was about 1am and when it got light Bill, a man I was working with and I went over there. The devastation was terrible and we couldn't find anything out but learned later the awful truth.

Later that day I was taken home and of course my parents were devastated. I didn't have to go back to Marconi's I was allowed to stay home and was released from war work.

On 29 December 1944 my sister was buried in a communal grave in Chelmsford cemetery with 18 of the other who were killed (they were unable to identify each body). There was a lovely ceremony conducted by the Bishop of Chelmsford. On the 2nd anniversary a memorial garden was dedicated by the Provost.'

Eileen Carter (nee Evans)

Dad's Army or 211 (101 Essex Home Guard) Z AA Battery, RA

We raw recruits were being given arms drill by Sgt X in front of Marconi House. Our first lesson was to stay alert and to come to attention smartly on command. After a few WW1 Parade Ground 'SHUNs' followed by 'STAND EASY', he wandered a few steps in the direction of the platoon captain, only to swing round with another 'SHUN' to catch us sleeping. Unfortunately the second time he tried this, his false teeth shot out onto the parade ground, to the merriment of the troops and the total loss of discipline. This episode prompted the question, was it really like 'Dad's Army'?

Later some of us were posted to 'J' Company, with HQ in Mildmay Road where St Johns' Ambulance is now. Here we were taught 'battle drill' with a great assortment of weapons, such as – Browning automatic rifles, Sten guns, Blacker Bombards, spigot mortars, rifle grenades, hand grenades, sticky bombs, and plastic explosives.

The most frightening thing was going on parade in the pitch dark to guard the ammunition dump at the drill hall in Market Road. The problem was the fixed bayonet waving about just in front of your nose held by the rank in front. This worry was resolved when incendiary bombs hit the dump, small arms ammunition went off for hours afterwards, like a firework display. The next most worrying thing was the sausage cooked for breakfast while under canvas at weekend training camps. I didn't fancy any, but everyone else came out in spots and some missed the Sunday parade.

In late 1942, probably as a decision resulting from the bombing of Marconi's on the morning of May 9th 1941, there was a dramatic development, which has not had much publicity. One evening I was watching a film at the Select cinema, suddenly there was an alarming hissing roar from outside. As it quickly died away it was obviously going up, not coming down. Later I learnt that it was the Chelmsford Home Guard Anti-Aircraft 'Z' Rocket Battery in action for the first time. On the way home afterwards I saw a German plane caught in a crisscross of searchlight beams towards Dunmow. As I watched, that part of the sky erupted in a cube of fire about 400 yards in extent, it was quite shattering in effect. I later heard that the another Home Guard 'Z' Battery shared the credit with an RAF night-fighter.

Soon afterwards the rest of the Chelmsford Home Guard was reorganised and many Marconi men then spent one night in eight on duty at the new 'Z' Rocket Battery in Central Park. Initial training took place at 185 London Road, followed by live firing out to sea at Walton on the Naze.

In Central Park there were 64 twin rocket projectors, organised in four troops Able, Baker, Charlie and Dog, located on the open area bounded by the river, the lake, Park Road and the then cattle market. Two GL radar cabins operated by ATS girls were sited on the other side of the river in the cricket ground. These cabins were linked to our ops room near Park Road.

Each projector could fire two 3' anti-aircraft rockets having a maximum altitude of 19,000 ft and a ground range of 10,000 yards (5.7 miles).

The heavy finned rockets were about six feet long and each had an adjustable nose fuse to be set to explode the warhead at the correct altitude. Two men manned each projector. The commands for altitude, bearing, elevation, loading, etc came over a sound-powered intercom from the operations room to a headphone worn by No 1, who relayed the orders to No 2. Each man set a fuse, No 2 loaded the rockets onto their guide rails and pulled them down onto the electrical firing pins and then set the elevation wheel. The firing pins were connected via safety switches to a firing handle and a 6 volt dry battery. No 1 set the bearing and reported 'Able 3 ready'; on the command 'Fire', he depressed the firing handle. If the rockets misfired, we had to wait 20 minutes before unloading.

The flight path of a rocket is quite different to the simple maths curve governing the ballistics of ordinary AA guns. Our operations room used a plotting table, and a prediction device like a three dimensional slide rule, to give us fuse settings etc from the radar data fed in by the ATS teams. The 'slide-rule' scales were prepared from a series of 3' rocket flight tests optically tracked and recorded by a team of ordnance experts sent to the clear skies of the West Indies.

Area command was provided by a master control centre at Sandon that covered all the local AA batteries, including the heavy 3.7' guns down on the Meads and the Bofors along the A12. Defence of local airspace was shared with night fighters, so that quite often we could not engage, but the battery did go into action and claimed hits on several occasions, usually single intruders. One of these came down in flames at Great Leighs.

One night a wave of bombers came in over Clacton and followed the railway line to Chelmsford. Here they dropped flares and then incendiary bombs. The Battery went into action several times. One Home Guard was commended for dealing with an incendiary bomb that had hit his Rocket Dump. In Chelmsford there were many fires, I saw the Prison and adjacent houses burning. That night sixty Chelmsford people were killed, many injured and twelve buses destroyed. On duty the next night I saw small bomb craters in a line from the south east corner of the site and across the river towards the radar cabins. This could have been the time a bomb splinter put the Radar out of action and targets had to be engaged visually.

One HG member was renowned for his near miss. One night there were 126 rockets aimed and fired south but his two lonely rounds went north due to a 180º bearing error and just cleared the top of the Railway Station, but then, it was only Dad's Army.

When there were no enemy aircraft about, we slept in three large Nissen huts where the bowling green was. Site services were provided by regular army staff. Morning tea was made by them in large buckets marked T for each hut. The other buckets, marked P, were emptied down manholes behind the huts, where the present conveniences are located. A very good free supper and breakfast was provided in a canteen near the Operations Room, as were the occasional concerts given by the Blue Ramblers Dance Band. The supper was usually fish and chips, the breakfast porridge and egg, the great mystery was fried spam, was it fish or meat?

In 1944 the flying bombs came in too fast for us to load and fire, we had to leave them to the Mosquito and Typhoon fighters. The 'Z' Batteries had come to the end of their useful life. We had the standing down parade of the 6th Essex Battalion Home Guard on Sunday 3rd December 1944.

I would like to thank the following ex-Home Guards for their help in refining my recollections of the above events : H.J. Ecclestone, E. Cranston, R.H. Oddy, D. Lloyd, D.J. Amos, C.A. Poulton.

Peter Helsdon

The Secret Army – SOE, Auxiliary Units and VI's

At the start of World War Two a covert British Resistance Organisation was established known as 'Auxiliary Units' to become active in the event of a German invasion. This secret British resistance organisation were patrols of specially trained and equipped civilians instructed to 'stay behind' enemy lines following an invasion. Operating from secret hideouts, they were trained to carry out sabotage and guerrilla warfare against the German invaders. These resistance fighters were considered completely expendable, with instructions to cause as much havoc and destruction as possible and their life expensctancy was considered to be a few weeks at best.

At the start of war Auxiliary Unit patrols were formed throughout Britain and had a peak strength of more than 1,500. They remained operational until 18th November 1944 when, with all threat of an invasion removed, they were stood down.

They were organised into local cells or patrols of 6-8 men with good local knowledge who could blend into the landscape and be useful in a tight corner. The units spent the summer and autumn of 1941 preparing underground bunkers where a supply of arms and supplies was kept. Great ingenuity was employed to conceal the locations, which varied from cell to cell, sometimes in the woods, abandoned dumps, out-houses, tunnels or underground bunkers. A local group of cells in Chelmsford, Wickham Bishops, Hatfield Peverel, Terling and Boreham was commanded by Captain Keith Seabrook, a farmer from Little Leighs.

These groups became part of a secret network of volunteer men and women prepared to be Britain's last ditch line of defence during World War Two. The Auxiliary Units were the first such organisation of its kind in existence in Europe. Having had the advantage of seeing the fall of several Continental nations, Britain was the only country during the war that was able to create such a resistance movement in advance of an invasion.

The formation of the units was executed using utmost secrecy and not even the unit members wives were meant to know about its operation. This secrecy would be fiercely protected during the existence of the units and after their stand down in 1944. It was not until David Lampe finally published his book 'The Last Ditch' in the 1960s that the true structure and objectives of the Auxiliary Units were finally understood.

The High Command HQ was located at Coleshill House near Swindon and this is where all unit members underwent intensive training. The Chelmsford Patrol, whose bunker was in woodland adjoining Hanningfield Reservoir, was led by H.C. Berry, with the rank of Sergeant, later succeeded by W.T. Macnab, and comprised R.W. Bartle, R.N. Carter, B.C. Ager, A.G. Taylor and H.W. Pratley. With the exception of Pratley and Taylor, they were all from the Test Department of Marconi's in New Street. In their twenties and thirties at the outbreak of war, and were undoubtedly recruited for their knowledge of radio and radar in addition to their other qualities. Pratley was in Marconi Research at Great Baddow and Taylor was with LNER (as it then was) at Liverpool Street Station. Brave men who were all prepared to fight to the last.

The British Special Operations Executive (SOE) was a World War II organisation officially formed by Prime Minister Winston Churchill and Minister of Economic Warfare Hugh Dalton on 22nd July 1940, to conduct guerrilla warfare against the Axis powers and to instruct and aid local resistance movements. On its formation, it was ordered by Churchill to 'set Europe ablaze'.

Its mission was to encourage and facilitate espionage, sabotage and reconnaissance behind enemy lines. In its early days it was also to serve as the core of the Auxiliary Units, the British Resistance Movement which would act in case of a German invasion of Britain.

One of SOE's most vital tools was clandestine radio communications. A special wireless set weighing less than 40 lbs. was developed. It looked like an ordinary suitcase. The Suitcase Wireless Set 'A' Mk 1 was one of the first wireless sets disguised as a suitcase to be produced by the Special Operations Executive (SOE) during the Second World War. It was introduced in late 1941 or early 1942 and its primary use was by clandestine agents working in enemy occupied territory. The plan was for every SOE organiser to take to occupied Europe a qualified Radio Operator. The design and construction of the transmitter and receiver is based on the British Army Wireless Set No. 18 Mk III which was, at that time, in full production and using readily-available components. In Louis Meulstee's book 'Wireless for the Warrior – Volume 4 Clandestine Radio', he attributes Marconi as the manufacturer of many of the British sets.

'When working in Test I was sent to Parson's Green to carry out mechanical and wiring checks on suitcase sets. For the wiring check we used a device called laughingly an 'Oxometer'. The test consisted of plugging connectors into valve sockets and the machine would circuit check automatically. Any faults found could then be rectified.

On returning to Parson's Green after a weekend home the place had been burnt down, the building being an old wooden Barracks.
On my return to New Street I was sent to War Office test under a Mr Jacobi, where the sets were tested. They came valved up with EF50s, the cases themselves were made up of receiver, transmitter, and a multipurpose power supply, Morse key and aerial. The receivers etc were all put into long biscuit-type tins complete with lids. A later version had a separate case with a hand-cranked battery charger.

I remember working all over Christmas getting the last consignment ready for despatch.'

Frank Whybrow

'The *suitcase set* comes in quite a lot of different versions, the three most popular being:

a) Type 3 Mark I or B3: Probably used 1941-43. 24" long suitcase, gross weight 42lbs.
b) Type 3 Mark II or B2: Probably introduced 1942-43 as a lighter version of (a). About 18" long suitcase. Gross weight 32lbs

Both (a) and (b) above are said to have been manufactured at the Special Operations Executive's factory at Stoneley Park North London. 32-42lbs is a lot of weight to carry but was about par for the course for a robust portable military transceiver in those days. The army manpack sets, Nos 18 and 22 were both in that range and carried as backpacks.

Carrying that sort of weight in a suitcase, to be swung nonchalantly in one hand while strolling down the street in front of the Gestapo, must have needed some practice and intestinal fortitude.

c)   A Mark III or A3: More of an attaché case size. Weight said to be only 8lbs but I'm not sure I can believe that. It appeared in the field in 1944. There is some documentary evidence at the Signals Museum that more than 400 of these were made by Marconi. I guess these were the ones tested at Parsons Green.

d) A Special version of the Army 18 set made for the SOE 'Jedbergh' teams. These were three man SOE teams consisting of one Brit, one American and one Frenchman, who started parachuting into France a little before D Day to help the various French resistance groups form into proper fighting/sabotage groups. They were in uniform, unlike the SOE teams of agents. The standard 18 set was modified to be powered by a hand cranked generator, tripod mounted. Three satchels of equipment could be made into one backpack, weighing 45lbs.

The technical spec for all of the above was much the same: Crystal controlled. Frequency range about 3.5 – 12 (or 15) MHz. LT or AC operation. At the lower frequencies, the range is said to be up to 500 miles with the right frequency selection but I suspect that the UK base stations all had propagation advice from somewhere and they dictated the frequencies to be used by the agents at different times and seasons. I guess they were all about 30 watts output to an inefficient aerial, said to be about 70 ft.

There was also a US Army set, circa 1945, called the AN/PRC1 Suitcase Radio. Two bands switchable 2-5 and 5-12 mc/s. 30 watt. Crystal controlled. 18" x 13¼" x 7¼". Weight 32lbs. Don't know where it was used.'

Bill Meehan

The Voluntary interceptor service was operated by the Radio Security Service during World War 2. The highly secret organisation was initially controlled by MI5 and was developed to monitor enemy wireless traffic.

How I Was Recruited into the Radio Security Service

'In the summer of 1941, when I was 19 years old, I was working for Marconi's W.T.Co. My job there was to diagnose electrical faults in radio receivers intended for use by the R.A.F. in their bomber aircraft. The receiver was known as the R1155.

Also working at Marconi's there was a pre-war amateur radio operator, Eric Taylor, G3FK, with whom I became very friendly. We had many talks about amateur radio and how I hoped to be able to obtain a transmitting licence when the war was over. At that time I was in the Home Guard, and how I hated being marched up and down the road, shouted at and even at times, sworn at. The only thing that I could do better, than even the NCOs, was to be able to send and receive Morse code at a few words per minute. I mentioned this to my friend Eric at one time, but then forgot about it.

One Sunday night the Home Guard had an 'exercise', although what this was about I had no idea. All I can remember was that it started about 10.00pm, it was pitch black (no lights anywhere of course) and it was pouring with rain. Most of the time I was wriggling about flat on my stomach with a stupid Sten gun on my back (no bullets) in wet grass, mud and dirty puddles. This, and the rain, went on all night until 6.00am Monday morning, when I was allowed to go home. Never was told what the exercise was about.

After getting home and having a bath to get rid of the mud, which had penetrated my Home Guard uniform and made me stink a bit, by the time I was dressed in clean clothes, there was no time for breakfast as I had to be at work by 8.00am. So, by the time I got to work I was in a right mood!

My friend Eric asked me what was the matter – and I told him! My language was unrepeatable as I told him what I thought of the Home Guard, the weather, the blackout, the exercise AND I WAS STARVING HAVING HAD NO BLOODY BREAKFAST!

Well, he calmed me down and we went to the canteen to get something to eat. He asked me if I could still do Morse, and I said that could only read at about a few words per minute. I got no answers to any of my questions.

Anyway, he brought in to work a Morse key and a home-built oscillator, and using a pair of Marconi headphones, he started to send me text from a daily paper. He instructed me to write down everything that I heard in block capital letters. (To make sure that I wasn't guessing the text he deliberately misspelled some words – sneaky!) Within a few weeks I was surprised to find that I was accurately copying up to around 20 words per minute. Naturally, I asked him what all this was for, but still got no answers. All I got from him was, 'Wait and see'.

Some weeks later, on a Saturday morning, there was a gentleman (rolled up umbrella, bowler hat and dark suit) at my front door, asking to see me - alone. 'What the hell is this about?' I thought, and I knew that my parents were concerned too. Well, we went into the Parlour, he shut the door and then he put a sheet of paper on the table. 'Sign there!' he said, somewhat belligerently. I was taken aback, but being just a lad and faced with this very authoritative figure, I did just that – signed my name.

'Right', he said, 'you have signed the Official Secrets Act and nothing that we talk about now must leave this room!' 'You will say nothing to your parents, relatives or your friends. It is essential that no one hears anything about our talk.' To say that I felt scared is a bit of an understatement. This man was very intimidating! He asked me quite a few questions about my family and then shot at me 'What political party do you belong to?' I answered that being only 19, and so unable to vote, I had not joined any political party. Finally he said that he had finished and prepared to leave. I asked him if he would have a few words with my parents, because they would be worried about his visit. Then he said, in my hearing, 'Your son might soon be doing work of very great importance to this country!' Sheer panic!

The next day, Sunday, I went to work, and bearing in mind that I had been told that I could tell no one about what had been discussed the previous day, I just mentioned to Eric that I had had a visitor.

'Oh, don't worry about that. I expected that would happen, it's just a security check' was all I got from him. Still had no idea what it was all about. Nothing was mentioned again.

A few weeks later, when I had all but forgotten about the incident, the Postman delivered a big parcel for me. Inside there was a covering letter from 'PO Box 25, Barnet, Herts.' Included were 'Signals Heard' pads, Message forms, a whole sheet of postage stamps, some envelopes stamped 'SECRET' in red, some larger plain envelopes and a quantity of sticky labels printed with the address 'PO Box 25, Barnet, Herts.'

I was informed that I had been recruited by the Radio Security Service as a 'Voluntary Interceptor' (whatever that was!) and I would be on G.S. (General Search). I was to monitor between 7000 and 7500kc/s. and write down date, time, callsign used, frequency and signal report for any station that I copied. Now this sounded exciting! So that's how I became a VI. I continued to send in logs and messages until November 1947.'

Ray Fautley

Stan Church and his brother Vic joined Marconi's and both started on the New Street shop floor as Instrument Makers. As conscription was in action, he decided to choose which division of the Army he wanted to be in so he joined the Territorial Army (TA) at Stratford. They were called the 54th East Anglian Division of Signals. When war was declared, he and his Division were at Shorncliffe in Kent. They were in France by the end of September. This is Stan's story:

'This was the period called the Phoney War. They moved through France and ended up in Lille near the Belgian border. There, they maintained equipment. On May 10th, when war became active, his Regiment moved into Brussels. They were outflanked by the Germans and were trapped in a pocket. Space around them got tighter and tighter.

Despite not being in the front line, his Division was always ready if enemy troops broke through the front line. They were trained in using arms. The situation was getting perilous and they knew that they would have to evacuate. The French and Belgians were holding the line with them, but the Belgians packed up and surrendered, which left a big hole in the line. The Germans poured through the hole and it looked as though they were not going to get out.

It was decided by Churchill and the War Cabinet that they would evacuate as many troops as they could. They set in motion a fleet of ships, mainly warships such as destroyers, which were going to come over and take them over the Channel from the town of Dunkirk. They destroyed their vehicles, because if all the vehicles were taken to Dunkirk, the roads would have been very congested. They started marching towards Dunkirk in the morning and reached there by nightfall on the 27th of May. Dunkirk was easily visible because the oil reservoirs there had already been bombed and this created a huge plume of smoke that could be seen for miles. On the way to Dunkirk, they had to take shelter by diving into ditches to avoid the hordes of Luftwaffe aircraft overhead raining bombs on them. They spent the night and the next day on the beach, while the Navy tried to bring small boats in to evacuate them, but it was a slow process. They were bombed and strafed on the beach, but because they were on the beach, the troops could dig themselves shelter against the bombs that were constantly coming down. Stan made his way to the Destroyer and boarded it, and made they made their way back to England being careful to avoid the mines that had been planted in the Channel. They were greeted as heroes, even though they had been beaten and forced to give up all their equipment. The Germans had been planning this operation for a long time and their army was very well equipped, and the Territorial Army was only a small army who were not used to open warfare.

All their equipment except for rifles was left in France. After Dunkirk, preparations were made for a German invasion. Stan went with his Unit to Broadlands in Hampshire, and that was his Ddivision's Headquarters.

After joining this centre, he was asked to go for an interview to be seconded to the RAF. He went to Uxbridge for a test, passed it and was posted to the RAF at a place called Upper Hayford in Oxfordshire which was an operational training unit. Over there were Hansen and Hamden bombers. Their job was to maintain the equipment, both wireless and flare piles which were large beacon pipes which were switched on when the bombers had taken off or returned. Stan enjoyed this job as he said, 'A blue job's better than a brown job.'
He was again posted to his Unit in the Territorial Army, but was put on reserve and had to stay home because he was of more use making equipment back at Marconi's as an Instrument Maker because the Army had lost so much equipment.

During that time he worked on equipment which was known as 'bending the beam'. This was producing transmitters which sent out false radio signals to direct German bombers away from areas that were vulnerable to bombing. He also made radar for the Navy, of which the later versions played a big part in helping to win the Battle of the Atlantic. He also joined the Home Guard. He joined in the D-Company which covered Springfield. He did Fire Watching at night time to look out for incendiaries dropped in the air raids. He compares his experience in the Home Guard to the comedy series Dad's Army because of many similarities to the characters among his comrades.

The Territorial Headquarters were based near the Multi-Storey Car Park in Parkway. They used to have to mount guard at night, and they had a regular soldier there, a sergeant whose job was to instruct the Home Guard. Many of the Home Guard were soldiers who fought in World War One, but some had never been in the Army. He says there was a lot of aerial activity over Chelmsford, where Marconi was bombed, and Hoffmann's was hit three times. Stan said that the worst raids were in the latter part of the war, where the Germans deployed flying bombs called V1s or 'Doodlebugs'. These bombs terrified the public because they made a loud screeching sound as they flew through the air. The third attack on Hoffmann's was by a V2 rocket which caused much devastation.

There were three other raids just on Chelmsford itself. During that time, Stan was working nights at Marconi's and narrowly escaped being hit, as the four buildings next to Marconi's were destroyed at the same time. Marconi's was the next building in line, fortunately for Stan, the German pilot ran out of bombs. He said that there wasn't as much bombing on Chelmsford as there was on London, Coventry or Liverpool, but there was enough to cause heartache among the public. At that time, the Americans arrived and built Airfields all around Chelmsford, the nearest one being Boreham. They had two Airforces: the Eighth Air Force, a strategic Air Force, which had Liberator bombers and B-19 Flying Fortresses and the Ninth Air Force, a tactical Air Force which had twin engine B-25 and B-26 planes which had a shorter range. They were based here so that they could support the D-Day invasion. Stan said that the Americans got up to a lot of mischief such as getting into fights at Dance Halls and stealing local people's bikes. He said that there were many comical episodes that happened that made him and his comrade's laugh. One such episode concerned a rocket battery where twenty pairs of rockets were poised to counteract air raids. One night during an air raid, the sirens sounded and the 40 rockets were launched. 38 of them went towards the advancing Luftwaffe planes, but one unfortunate person had got their coordinates 180º in the opposite direction. Two rockets went flying in the opposite direction and were nowhere near their target. Another such mishap occurred in August. It was a hot night, so the Home Guard took off their clothes and got in the cooling ponds. Suddenly the air raid siren went off and the soldiers got out of the ponds equipped with boots and rifle, but nothing else! Stan said that being in the Home Guard was similar to being in Dad's Army because of all the hilarious mishaps such as this that occurred. At VE Day, there was a big celebration at his Mother's pub in Baddow Road. However, at VJ Day, after the atomic bombs were dropped on the two Japanese cities, Stan felt that warfare would never be the same again and felt quite disconcerted. He felt that winning the war opened a new chapter of the future. After the war, Stan went back to work at Marconi's.'

For Stan Church

'**Every** day we had bulletins with our Workers Playtime. During late April the rockets had all but ceased. The big event locally was the wedding of the girl next door. Coupons were scrounged for material for the dress. We used to go to the market on Saturdays and tell the stall holders we worked at Hoffmann's and we could buy clothes without coupons. We had twenty to last six months and a coat was eighteen so we cultivated our own Black Market very successfully. Anyway the wedding was a huge success. We kept chickens so six eggs went for the cake. The neighbours gave small amounts of butter and sugar. We sure wished John was around with the PX rations. Anyway a large cake was produced and we all tasted it and it was good.

End of April the weather was hot. I was then on night work so saw some of the lovely weather. All getting excited as the War seemed to be ending. Italy had been invaded and almost defeated. Most of the Army had surrendered. I think Dictator Mussolini had been captured and hung upside down from a pole. We were joyful.

May 7th - Fall of Berlin. Crazy, crazy time and on May 8th - Germany surrendered. I did not go to work that day and neither did Leslie. We caught a train to London and walked from Liverpool Street station right along past St. Pauls and bomb damaged City to the Strand, which was full of people singing and dancing down the Strand toward Trafalgar Square. I think what buses there were gave up the struggle.

The Square was just a mass of people with soldiers etc. sleeping in doorways and curbsides with their rifles propping up their kit bags. Everyone had seen the newsreels lately. Well I was there! We danced and sang and joined hands in a huge crocodile down to Whitehall and Mr Churchill and his Government came out on the Treasury Balcony to a huge roar of voices. He gave his Victory sign and we blew him kisses and laughed and shouted. London had never seen the like. The crocodiles of solid masses of people then wound their way through St. James's Park to Buckingham Palace. A huge chant began. 'We want the King' over and over until at last the balcony doors opened and out came the Royal Family. Another great roar of singing and laughing. By this time it was around 6 pm and the doors opened again and the Royal Family and Mr Churchill came out to cheers and roars from the crowds. To this day I still do not know how he could have got from Whitehall to the Palace but there he was. It was a blazing hot day. I cannot even remember eating or drinking at all. Spent all evening around the Palace. Must have looked good from inside.

It gradually became twilight and the Palace was floodlit for the first time. The balcony doors opened again to the roars of the crowd. Who ever imagined going home? We lived thirty miles from London. Midnight and Big Ben chimed and the Royal Family came out again for the last time.

Got back to Chelmsford around midday. A note from Leslie's Department saying 'Where was you?' and please report as soon as possible. I went back to work the next day and were all informed that the War was not over in the Far East. It was hard to settle down to work with the beautiful summer days outside.

I heard from sister that there was a Court Summons to make me return to the factory. As she was the only one who knew my address nothing was heard about it. The War in the Far East dragged on until August. I went out to buy a paper with screaming headlines 'Atom Bomb Dropped on Two Towns in Japan with hundreds of thousands killed.' How we cheered. Japan surrendered, the date I cannot remember. More celebrations around the Palace, this time we had somewhere to walk back to. The days went into weeks and everything settled down.

I bought bits of second hand furniture as all new furniture was controlled by dockets and you could only qualify if you had children and were bombed out. We managed a bed, table and armchairs and brought them back in three journeys with a barrow. This seemed the end of my war. '

Olive Cox

'I spent a very happy 48 years with the Company retiring in 1987 one year early mainly due to health problems. In early May 1939 my father and I were interviewed in the old canteen by Jack Frost the then General Manager with a view to my employment. The next week I had a letter asking me to start work in the Standards Room under Mr. R. Cartwright on the 19th of May. For the first few weeks my pay was 13/3 per week.

I spent three years with that section and was mainly employed drift running and compensating Franklin Drives, Master Oscillators used in SWB8 and SWB10 and SWB11 transmitters. My next section was Mobile Test which was led by Bill Stroud testing quite a variety of transmitters used in portable stations. Whilst there, it was my luck to test the first VHF transmitters for vehicles. About 1950 work took me to VHF Development Group. Jobs were varied and most interesting here, most of the section subsequently moved to Writtle after a year or two.

There was a shortage of work in 1958 at Writtle and so I was seconded to Rivenhall to help with a large project going on there. Two weeks after my marriage in 1961 I was recalled to Writtle and give the task of forming a section to Repair and look after all the Test Equipment on the site. This was an enjoyable time and the help given me by permanent staff and apprentices was much appreciated. We moved to New Street in 1974 and finally to West Way from where I retired still repairing Test Equipment although more specialised.'

Kenneth Hutley

'I was the development engineer responsible for the production versions of the SWB8X and the SWB11X. As it happens I was the first Engineer to move in Building 46 in 1950 and my first job was the uprating of the two SWB transmitters to the X edition to meet the new post war 1947 CCIR regulations and to incorporate the use of Single Sideband voice transmission.

Not many new ones were produced, in fact we had a contract to upgrade a lot of the old War Office stock and I remember going to various MoD Depots to pick suitable models to uprate! This was because restrictions had been placed on the purchase of new equipment, but the refurbishment of old equipment was allowed (even if it cost twice as much!!)

I well remember A.J.G. Corbett and the preparations for the Gothic installation. Also I seem to remember that the use of the Worldspan Transmitter was its first operational use and there was a great deal of panic in the lab (Building 29 then) to get it fully up to performance for the Gothic.'

Bill Barbone

'

I joined Marconi's Test Department from the Chelmsford Tech in 1943 and leaving in 1974, my memory of those times is hazy but I think Mr Robb, who had an office in the front building, was the Chief of Test.

As a junior, I started with the boring job of testing components like capacitors, inductances, Muirhead drives and coils. The area for testing coils was at the end of the girls' winding shop upstairs. A dangerous place for self-conscious young men! Later I moved on to High Power Test. This was the 'big bang' area – before Safety at Work rules and regulations! SWB 8s and SWB 11s for the Admiralty were a fairly steady work load and later, high power Broadcast Transmitters.

High Power Test was run by Mr Whiteman and later by V.J. Sandy. Among the 'oldies' were G C Baker, and Mr Wylie – who I think suffered as a Marine Operator earlier in the war. Then there was Alfie Amos,

Tommy Tomlinson, Doug Hills, Ecclestone and many others whose names have slipped my memory. G C Baker would regale us with the excitement and problems of taking his nanny goat(s) on the back seat of his open Riley car to visit Sandy's goat the previous evening.

We were handling lots of kilowatts but I don't recall anyone getting seriously hurt: even a young man known as 'Fearless Fearman' who insisted that power was off at the wall and proceeded to put a screwdriver across the terminals of a SWB 8 transmitter to prove it. Unfortunately he was very wrong, but lived to explain loudly that it was somebody else's fault!

In the exposed corner by the entry to the High Power Test was a lash-up to provide very high voltages (like lightning!) mainly for testing or proving high voltage insulators. Very, very high voltages were generated (enough to make your hair stand on end!) and it took a bit of know-how to set up the tuned circuits to get the required voltage. Wylie was good at this.

The transmitters were built in the workshop below Marconi House opposite the canteen. I cannot immediately recall the name of the Foreman, but the charge hand was Charlie Pashley. I think he had been involved in the Altmark event off Norway in the Second World War and survived. I got to know many of the fitters. They were a good and friendly lot, some of whom later worked with me on overseas installations. At the entrance to Marconi House (or just behind) there was an entrance to the lower ground area or cellar where the archives and records were kept in the care of an old RN sailor who took part in the Zeebrugge attack in the First World War. He certainly had several fingers missing. I can't think what I was doing there but he was always good for an interesting chat. Another link to the First World War was Diggens who was a sort of general helper around Power Test. He would cheerfully relate the most unpleasant experiences of fighting in the trenches.

In due course I went into the services for a spell and on my return became involved on overseas installation work, but that's another story.'

Tom Gutteridge

'**We** celebrated the end of the War with much jubilation with hundreds of people dancing and drinking in Tindal Square. We managed to make cakes, jellies and sandwiches from our rations. We dismantled the Morrison shelter and the steel top was placed on the road and a fire lit around which we danced. The next day we found the heat had melted the tarmac and the shelter top had stuck to the road!'

Stephen Williams

There was an unusual sequel to the attack on the New Street Works. On 12th June 1945, officers of C Squadron, 8501 Wing, R.A.F., examining a fire-gutted photographic building on Quedlingburg Airfield in Germany, found a scale wooden model of the Marconi and Hoffmann Works, evidently prepared from aerial photographs made by the German Air Force during the war to help their aircrews to find New Street and bomb it accurately. The model was retrieved and presented to Marconi's by the R.A.F. officers. For a long time it stood in the main entrance hall at Marconi House, Chelmsford, near the Roll of Honour of those who lost their lives.

They shall grow not old, as we that are left grow old:
Age shall not weary them, nor the years condemn.
At the going down of the sun and in the morning
We will remember them.

Lest we forget.

Some Seventy years after the end of the war - The Marconi water Tower still carries its wartime camouflage as a church with 'stained glass windows', while the railway sidings and nearby buildings still clearly show their disruptive colour patterned paint- especially clear after a rain shower.

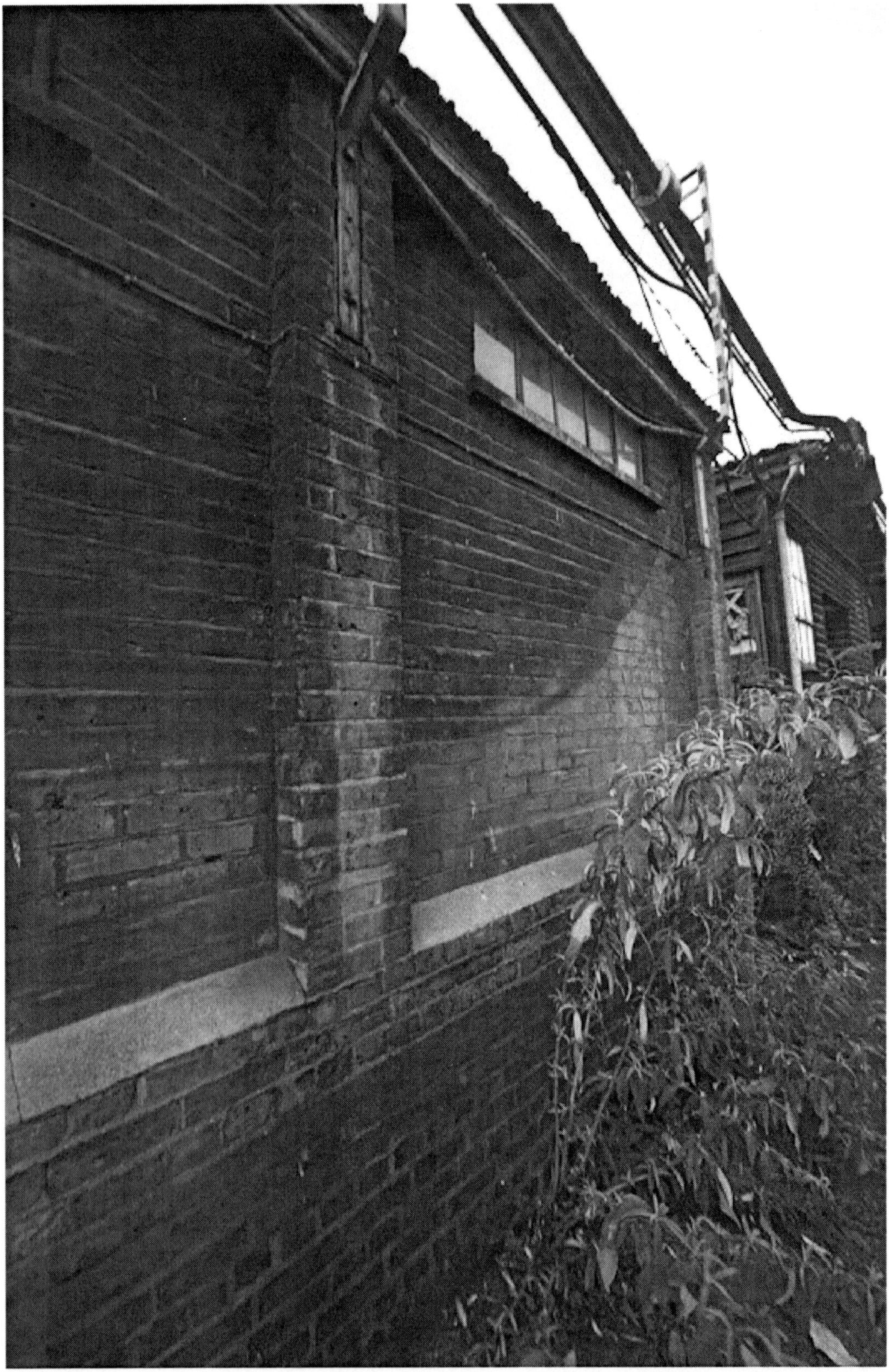

CHAPTER TWELVE

After the War was Won

After the war, returning to peacetime activities, the great post-war expansion of business severely overloaded the New Street factory and offices.

The foundations of the Marconi Radar company had actually started in a remote corner of the New Street Works with the Vacuum Laboratory created in 1929 to aid the ongoing development of thermionic valves. The lab was run by G.M Wright, F.M.G Murphy was one of the first engineers and he was assisted by glass blower E.G. Heriott. In October 1932, the Manager was Dr George M. Brett who described the Laboratory.

> 'The vacuum laboratory as I took it over was a single room, maybe 20ft by 25ft which contained a diffusion pump outfit, glass bench, spot welder, hydrogen furnace, my desk and little else. The place was gloomy, claustrophobic and squalid and I thought a poor exchange for the HH Wills Physics Lab which I had just left.'

After the War the Company organised a new Division to handle the new radar business and in 1968 this moved from New Street to the empty Writtle Road Works. The war-time valve factory in Waterhouse Lane became the English Electric Valve Company Ltd in 1947.

Some of the other activities that had been started at New Street also had to be moved out. The Aeronautical Group went to the Basildon Factory in 1954, the Broadcasting and Television Development went to Waterhouse Lane in the same year and the Marine Company moved to Westway, Chelmsford, in 1963.

New technologies after the war gave rise to further new Divisions, including Line and Space and Broadcasting Division. Many of the large dish aerials that can be seen up and down the country and abroad, have been provided by Marconi, including several of those at the Goonhilly (Cornwall) station of British Telecom.

> '**Around** 1945 or 1947...I was living at Tiptree then, which had no direct connections with Chelmsford at all, Chelmsford was somewhere I'd never been to in my life . . . as a young lad, it was the other side of the world . . . I was catching a bus . . . just after 6, a Moore's bus . . .running from Kelvedon and when I was going to night school . . . nine o'clock bus from Chelmsford to get home at 10 if I was lucky having cycled all the way uphill from Kelvedon to Tiptree, and how I did it for six years and stuck it without getting fed up with it I don't know to this day really . . . I can still remember catching that early morning bus which was largely full of Marconi and Hoffmann workers. Hoffmann workers always had their peculiar smell of cutting oil and on a damp morning, when you got on the bus and there was a smell of steaming Hoffmann workers, cutting oil, all smoking Old Holborn tobacco, I remember that smell very vividly.'
>
> *Derek Pope*

> 'I think one of the things about being locked in this building [New Street] - it was another world in there in the winter you seemed to go in and it was dark and you came out and it was dark but the one thing which used to liven up the day was music while you work, not enough's been said about this . . . half and hour in the morning and half an hour in the afternoon ..it was something you really looked forward to... and if the tempo was fast enough you, piece work jobs, your jobs, you hammered away twice as quick in fact when some of the larger bands appeared down the corn exchange, we gave them the tunes we would like them to play but we were told that they were limited to what tempo they were allowed to play because of accidents that could happen.'
>
> *David Newman*

'**Another** memory I have is of a gentleman who used to walk through the workshops mid morning and mid afternoon spraying what I suppose was disinfectant or something up into the atmosphere which was a very highly perfumed, scented thing and you even got to look forward to him coming round because it broke the day up a bit.'

Derek Pope

'I think one of my outstanding memories is the automatic machine, lathes and things that were clattering, making nuts and bolts and all different things, but the noise as you walked past was quite horrendous and as a young sprog apprentice I can remember being quite frightened almost by the sheer volume of noise and dangerousness of the machines as they appeared to be at that time nobody was every working them, they were all automatically clattering away . . . it was a bit of a black hole where they worked. '

Derek Pope

Forty seven years with Marconi's

'I had the ambition to follow my father into his retail grocery business, but he recommended that due to the extra work of rationing and collecting coupons, it would be advisable for me to seek an alternative career which might be more beneficial and secure. With the help of our neighbour Bill Pannett, Deputy Manager in the IDO at New Street, he was able to find a position for me. So on Monday, 4th January 1945, I started work as Office Boy under the guidance of IDO Office Manager, Charles Price (father of Doug Price, ex Radar Company).

By October I was transferred to the Wiring Shop, managed by George Stock and Cecil Myhill, to start my practical career, under Chargehand Jimmy Cowburn. On this day appeared for the first time in the Daily Mirror a cartoon character called 'Jimpy' and as I must have resembled him in many ways, someone gave me that nickname and to this day, those unaware of my Christian name still call me Jimpy.

That Christmas MWTCo Managing Director, Admiral Grant, came from London to address his 'Ships Company': standing on a dais in the Instrument Shop he thanked the workforce for their hard work and to thunderous applause announced that everyone would receive a Christmas bonus. I received an extra shilling!!

In October 1946 the first Marconi Apprentice Training Scheme was launched and I commenced my apprenticeship as an Instrument Maker at Pottery Lane, working in a variety of Nissen huts, a legacy of the war years. Mr Hitchens, the Training Manager at New Street, oversaw Mr Hillman and his team of instructors, gleaned from various areas within Marconi's, such as Ted Cordery – Wiring Assembly, Charlie ('File that flat for me'!) Sweetman – Instrument Making, Bob Thrift – Lathes, George Snow – Mills, John Cooper – Turrets, and Jack Whittaker – Sheet Metal.

That winter was a reminder of the contract associated with apprenticeships. Whereas workshops at New Street closed down when employees couldn't get to work due to heavy falls of snow, apprentices were expected at their daily workplace and I had to make my way from Great Baddow to Pottery Lane on foot. On one occasion John Cooper gave me a lift on his motor cycle from Longfield Road off Baddow Road (where John still lives), arriving two hours later, and then having to thaw out with the help of coke braziers – the only heating in the huts. John Atkins (son of Les Atkins, Tool Room Foreman), Bill 'Baldy' Brewer, Cyril Chorley, Alan Hart, Colin 'Dingle' Humphries, Tony Martin, Dennis Martin, Les Sayer, and Derek Tiffen (later to shine with Chelmsford City Football Club) were part of that first intake.

At the end of our first instructional year we went our various ways. I spent my second year in the Instrument Shop at New Street under Charlie Britten with Jack Stock as chargehand. In October 1948 I went to the Transmitter Development Labs in Building 46 under Mr. Burdick, Frank Bishop and Frank Newton, working directly with engineer Dennis Hart.

My final move was to join R & D Workshops at Great Baddow in the Instrument/Wiring Assembly Department under Superintendent Bill Bush and Chargehand Fred Leach. I reached twenty one at the end of my five year apprenticeship and moved into the skilled wage bracket, doubling my apprentice wage. Within a week I was back to £1 a week as I was off to do National Service in the RAF. In 1953 after demobilisation I returned to Baddow Workshop and the Assembly Shop with Gordon Denney now Superintendent. In 1957 the opportunity arose for me to join R & D Planning Department under Office Manager Bill Clayden, a staff position and pensionable! 1960 I became responsible for the planning in the Assembly Shop under the foremanship of Fred Leach. By 1963 George Williams, now Workshop Superintendent, had been asked by Ray Stiles, Manager, R & D Workshops, to nominate a suitable candidate to join a newly formed R & D Estimating Department. I accepted the challenge and after a stiff interview with Company Chief Estimator Harry Ashworth, found myself travelling every working day for twelve weeks on the Company transport in the capable hands of ex bus driver Clarrie between Basildon and New Street, to learn basic estimating techniques in the Basildon Estimating Department.

In October 1963 four rookie estimators formed the R & D team under Works Estimator Ron Aston, situated in Harry Ashworth's office area on the fourth floor of Marconi House, then to offices overlooking Section 120 underneath the canteen and finally a permanent base on the second floor of St Mary's House, Victoria Road, all the time recruiting and training new staff. In 1972 the Group, having covered all aspects of work associated with R&D Workshops at Building 29, Great Baddow, WHL and Writtle, was disbanded in a policy change and split down the middle with half the staff joining Communications Estimating Department at New Street and the rest of us setting up a new Radar Estimating Department at Writtle Road, under newly appointed Chief Estimator Les Biggs. Here again, from a handful of estimators in a small office in E Block we grew into a formidable office of sixteen experienced personnel based on the mezzanine floor overlooking the Chelmsford to London railway line. For the next twenty years I was involved with all aspects of manufacturing, budgetary and MOD technical costing on a wide range of radar and associated equipment As an aside, I was seconded to the Training Department together with John Benbow, Planning Manager, to head up and organise a first year Production Introduction Programme for GEC sponsored students studying at Bath and Bradford Universities and Portsmouth Polytechnic, to attend a twelve week course during their summer recess at Writtle Road. Their second rear summer course was spent in the commercial areas and third year in engineering. Throughout they were allotted engineering tutors: Chris Bousfield, Harry Fancy, Clive Gildersleeve, Paul Hibbert, Brian Partridge and Frank Savill, to guide them at Writtle Road. They wrote up an extensive thesis on their three years in industry and together with their academic successes achieved a Masters Engineering Degree. This involvement under industrial conditions was an ideal way of achieving an understanding of what to expect when they joined the Company after University as junior engineers. The programme continued for six years with a new intake each year.

During that time Estimating Manager Les Biggs was awarded an MBE in recognition of his liaison work between Radar Company and MOD Technical Costing Department, an award richly deserved and an honour for office personnel who supported him. By November 1992, I had the opportunity to retire after 47 years of interesting and rewarding association with various sections of the Company, an experience I couldn't have enjoyed if I had gone into retail grocery!'

Ron Hurrell
ex Radar Division Estimating Department

In a photograph taken in the old Instrument Shop New Street at Christmas 1946 Admiral Grant, the then Managing Director, addressed the assembled factory staff whom he called his 'Ships Company'. The Mayor of Chelmsford was in attendance and as far as we can gather Alderman Sidney C. Taylor who owned the Essex Weekly News and The Lord Mayor of London were also present.

'**Being** that the Instrument or the Machine Shop was the largest area, there were occasions when, when English Electric took over Marconi's they put up this staging in the Instrument shop and Admiral Grant the

Managing Director came along and referred to us as 'shipmates'..or 'My ship's company'...he introduced us to Lord Nelson of Stafford who then in turn introduced us to his son and the cry went up 'half Nelson'.'

David Newman and Derek Pope

A Personal View of Marconi over 42 years from 1946-1988

'**My** look back at the mixture of individuals who, in spite of minor disasters, somehow or other helped to turn Marconi's into a major Company. Many would not have got past the modern style of Personnel Department or fitted any mould but somehow gelled together to achieve results. As the biggest misfit of all, I journeyed through the Company like the proverbial Jew leaving a few signatures on test specs and drawings and being remembered by the people I sheltered with on the way for staying too long. I was lucky, I survived, and came out the other end, unlike Fred Piper and Ray Weatherhead and countless others who fell by the wayside.

After serving in the Fleet Air Arm as a Radar Mechanic, I joined Marconi via the College and an interview with J.P. Wykes the Chief of Test in his office in the front of the old building in New Street. I was successful and ended up in Test Equipment in Building 29 under Joe Chamberlain. That's the way I joined the Company, so now on to the people I worked with and all the other friendly people who helped me over the years.

J.P. Wykes could be hell to work for but, could be forgiven as he would never allow outside criticism of his staff. He was extremely helpful when anyone had a personal problem. Test Equipment staff were a bunch of idealist radicals, led by Freddy Roberts who, later became Labour Mayor of Chelmsford, at a time when Russia was considered to be a model society and we all believed we would live in a great new world. Mr. Tattersall was a gentleman of the old school who sported a twirly ended moustache and impeccable manners which were applied to everyone. He was at sea when the *Titanic* sent its S.O.S. message and the message he received, sent the ship he was on steaming towards the area. Lady Telford, who was then Betty Shelley, provided the clerical backup.

Who remembers the winter of 46-47 when we survived with no heating and a few portable generators in the yard to provide a limited power supply? Were you in the main works when Admiral Grant assembled his 'ships company' to announce the takeover by English Electric? How many of you remember Miss Partridge, a tester in the Standards Room, who rode a 250cc motor bike, smoked a pipe, liked a pint with the lads but still managed to remain feminine? Did anyone know her Christian name, as I never did?

I was sent from Test Equipment to Aircraft Test in the old Skating Rink in London Road with its coke stoves and no windows, where in winter, you went in when it was dark and were surprised to see that it was snowing at lunchtime. Alternatively, did I turn up at the palatial establishment at Pottery Lane which I remember as a group of Nissen huts somewhere behind Mr. Webbers Browning Meadow along the Broomfield Road. George Barratt was the Shop Foreman here and his team assembled the English Electric television set, a post war Marine Receiver and an Echo Sounder. Design of the television set had been done in the Research Labs at Baddow. Did Barry Sitch and I tune the receiver module of the television as fast as we could so we could pull George's leg about not being able to keep up. Ken Paisley did the overall testing and when we ran out of work we would watch Andy Pandy and the Flower Pot Men in the afternoon as few people had even seen television at that time. Reg Bowler supervised the rest of testing and had a small office labelled 'Chief of Test'. This survived until Mr. Wykes arrived when he was told to take it down because 'there's only one Jesus Christ here'.

At the Skating Rink aircraft equipment was assembled by Ben Burnell with his Chargehand Tel Harris and a very experienced crew including Jack and Bernard who worked together for so long that, like an old married couple, they communicated without talking. I joined Reg Norfolk's team which had a square loop aerial. This was fitted within the fuselage of an aircraft. The square window in the aircraft cracking, was the cause of the early Comet Aircraft crashes. My next job was preparing for the testing of the prototypes of 'Green Satin' which was the first of the generation of Doppler Navigators.

I worked with the Development Group at Writtle learning about the equipment, writing test specifications, liaising with Test Equipment on making the necessary jigs. Ask Bill Claydon about the trials and tribulations of producing those first few prototypes. How was it that a state of the art equipment was developed in a wooden hut on a site that regularly flooded in the winter and, where transport for most of the staff involved was their cycles? Testing of the prototypes was done in the annexe at the Rink. Charles Griffths didn't seem to turn a hair when a heavy steel can was blown off the Transmitter Unit on pressure test, hitting the sloping ceiling and hitting the deck; it would have caused a nasty accident if it had hit anyone. At this juncture Aircraft Division went to Basildon and Green Satin production was completed at New Street, the testing being done in the old canteen. You will all remember very tall electrician Bill Clements and his five foot nothing mate who carried out the necessary wiring.

Next was Receiver Test with a Microwave Television Link, designed by RCA, handed over to Microwave Development Section at Writtle with an embargo on any changes. This must rate as the worst ever equipment produced by the Company and when it was sold to the Russians for 625 line working, when it was designed for 525 line working this generated a few problems. Do you remember the Outside Broadcast Vehicle with Russian markings which stood in the yard for several weeks? Taffe Dawson reminded me the last time we met that he was the apprentice helping me on the Baddow Tower when we carried out the tests between there and the Radar masts at Bromley. 30 mile Range tests from Bromley to the Water Tower at Aldeburgh were a complete fiasco due to a very low bank of hills. Being stuck on the top of a Water Tower for two days in the winter when you're sure within an hour or so that there is no signal, doesn't improve your temper. Do you remember Receiver Test before the Perimeter Cage was fitted where people built 'hidey holes' behind pinned up drawings?

Mr. Wykes had moved on to become Manager of Maritime Division and he offered me a job in Technical Sales. Here I met Dick Rocker and George Brunton and through him ships installation engineer Dicky Cann. The latter two were a pleasure to work and drink with and were known and welcomed in every shipyard and on every ship I visited. At the end of a year I complained that I hadn't enough to do and was loaned to Maritime Development. At the end of three years I was wiped off the Divisions books but as Development wasn't told I belonged to nobody.

I worked with Charles Burnham on the Naval version of the Wideband Amplifier which was used on ICS 2 and 3 and eventually sold to America for their Helicopter Landing Ships. At this time Jimmy Gould developed the 1500 Synthesiser which was used on many equipment's and must hold the record for sales of any design. There was some field-work and I once flew from London to Glasgow with Charles Burnham and travelled on a bus to Greenock accompanied by a drunken Scotsman who played his bagpipes. When he paused he berated the two Sassenach's who dared travel on his bus. We stayed in an old hotel in Gounrock, sharing a large double bedroom, lit by one ceiling light which was fitted with a single forty watt bulb. When we turned on the bedside lamps, the centre lamp went out. We were working on the conversion of wartime frigates to weather ships at Greenock Shipbuilders and eventually we sailed on the 5am tide and returned at 9pm. A very long day, but a sail down the Clyde Valley and round Ailsa Crag on a beautiful sunny day is something everybody should experience at least once in their lifetime.

I could carry on but getting back to more mundane things Maritime Development was split with the commercial side going to the Marine Co at Widford and the Defence Comms moving to Billericay. I think this was to squeeze in the first printed circuit unit into New Street at the top of the yard. I went with Defence Comms to Billericay where we stayed for about 2years. There was a big fuss about moving but nothing like the fuss that arose when we moved back: proving that there's nowt so strange as folks. Memories of Billericay were Steak and Kidney pie and bar billiards in the Red Lion, cheap shirts from the adjacent shirt factory and travelling across Galleywood Common in Tony Evan's old Ford on dark snowy nights with the windscreen wipers driven or not driven from the manifold and with its one candlepower headlights.

Waterhouse Lane must have been our destination and I was here for several years during which time I became more involved in Post Design Support. This involved solving problems on ICS 2 at Widford and Wembley and also engineering support to Post Design Services at Widford. This period is rather fluffy as I had so many masters and so many fingers in different pies. Where was Defence Comms section when the ETG was developed? Would I be right in thinking we had a berth in the Radar Company for a while? ETG was interesting as we mounted a 30ft transmitter whip on top of a large fin cooled casting, Ron Sewell did the casting design and we ended up with a good looking unit which was easy to mount. I had a week in Holland at Vlissingen (Flushing) which was a combined seaside town, Ferry-Port with a shipbuilding yard when the Dutch frigates were being set to work. An enjoyable week as the shipyard closed at 5pm leaving Martin, my colleague, and I to have a pleasant evening meal in a small moated town in one or other of the restaurants in the town square.

When Mr. Sosin became Manager, he insisted I should work for Dennis Hart in Building 46 so I had a new boss but carried on doing the same sort of work except that my visits now included Whitehall, the bunker at Northwood, with its patrolling armed guards with dogs, the Nimrod Airfield in Scotland and the Naval communication bases at Crimond in Scotland and another situated near Fleetwood. During this period Bob Nice took over as Manager and when Support Services PDS contracts expanded and required a full time engineer, I was allocated to Widford2 but was responsible to Bob Nice. Most of my work was then with the Navy and because ships were only available for very short periods and at short notice, I worked closely with the Fleet Liaison Officer, much to their annoyance who only knew that we were spending their money after a fixed arrangement had been made. However, we had Sid Johnson in the middle placating the Marconi Contract Manager and the MOD civil servants. He was a great peacemaker. A great time this with ship visits, days at sea, trips on submarines, and visits to aircraft carriers and nuclear submarines.

At the end of March 88 I retired and now looking back, realise that it couldn't have been like this, could it, we never all helped each other and all of us were not programmed to laugh when we should have cried. When I now think back and say I'm glad I joined, I must surely have finally lost my marbles.'

Roy Hubbard

'**The** apprentice master was Mr Hithcins, [a Yorkshire man] who wrote everything you told him down on a cigarette packet and then put it in the wastepaper basket and threw it away …I was in the shop two years, never moved out 'cos he lost the paperwork. Whenever you saw him he would say 'where are you now lad', or where are you working now? He never seemed to know where you were and what you were doing, and if you liked where you were you'd stay there a bit longer.

I remember he took me on my interview afternoon and he seemed to think that I could possibly progress to become a draughtsman, never thought of that, but he then took me up to see Mr Meade, who was the so-called chief draughtsman, who was quite an awe-inspiring little man I learnt afterwards he was a very nice little man . . . you know, a lot of us owe an awful lot of our career to people like him.

He was a very caring man but he was a little bit daunting to a young sixteen year old and asked me a lot of searching questions about why I thought I should aspire to this task of being a draughtsman, but yes I can remember . . . Hitchins saying if you stick it lad and get your ordinary national, even your higher national, there's no limit to where you can't go.

..latterly of course in Marconi's everybody was a manager, well not quite, but . . . he was not a manager, there weren't many managers about, but he did care for . . . if he had you in his office because you hadn't done well on day release he didn't want to tell you off, it was just that he was upset that you weren't doing as well as he hoped you would do, he actually meant it . . . you only appreciate these people afterwards

[In New Street] there were people in rooms, you didn't know quite what they did, all secret work . . .It was run by a Mr Green who'd been there since the year dot . . . lovely man, and I can remember one of his boffins who sat in a corner surrounded by a great knitting machine of wires, valves and god knows what else, and Mr Green went round, I happened to be passing, and he was sort of looking at what this chap was doing, and he was telling him in great details as to what he was experimenting with and everything else, after a while Mr Green sort of said but you were doing that last year. This was the tempo that these people obviously worked at . . .'

David Newman and Derek Pope.
Abridged from interview with Ray Murrells as part of a series done by the Marconi veterans for the Chelmsford Museums in 1998.

Reminiscences

'I too am a little diffident towards running the risk of boring your readers by reminiscing about my extended service with Marconi's Company – but you did ask for it!

I was demobbed from, the RAF in July 1946 having spent the previous five years as a pilot, touring the world at the country's expense. In that comparatively short space of time, I had visited Canada, The States, the British West Indies, North Africa, India, Ceylon and Burma. I also spent a few months in the United Kingdom! There were inevitably some dodgy moments but on the whole I thoroughly enjoyed my time in the RAF, but the freedom and lack of real responsibility came to an abrupt end when I was forced to begin searching for suitable employment.

For some months I was frustratingly unsuccessful. Ultimately a cousin, who was the assistant Works Manager, suggested I should approach his employers, Marconi's – to see if they could help. I was a little unhappy about possible suggestions of nepotism, but gratefully accepted an appointment in the Commercial Department, affectionately known as Jolly's Follies, in its Publicity Office. Commander Jolly, the Department Manager, was a prodigy of the then General Manager, Admiral Grant, who always referred to the Marconi staff as 'his ship's company'. He knew everyone by sight and most by name, but this very pleasant closeness was rapidly eroded a short time later when the English Electric Company came on the scene and we all became just numbers on the Company payroll. I was not impressed by spending my hours composing Press Releases about products I had never seen and had no idea what they were intended to do, and so when I was asked if I would be interested in a temporary two-week secondment to the Aeronautical Division, I jumped at the chance.

The Aeronautical Division, managed by L.A. Sweny, an ex-RNAS Commander, was scattered over New Street, Hackbridge and Croydon, both in Surrey, and Writtle. The latter was a small development establishment, mainly comprising a collection of World War 1 wooden huts with a modest modern factory, in a delightful rural setting. It's current task, when I arrived there, was the development of an innovative generation of miniature communication and direction-finding equipments, hopefully to be installed in the many postwar aircraft beginning to emerge from British and foreign factories no longer engaged in satisfying military requirements.

The Chief Engineer was Christopher Cockerell who later retired from the Company to concentrate on the development of the Hovercraft and becoming internationally famous as its Father. As I had only recently bade farewell to the Royal Air Force, this appointment enabled me to preserve a welcome continuity of contact with aeroplanes which had constituted a very large part of my life during the five previous years. In fact, many of my new working colleagues were ex-RAF and so our little community, comfortably housed in our old wooden huts in the wastes of wildest Essex, and openly exhibiting the RAF's light-hearted approach to life in general, was really like a Service unit in civvy clothes! Therefore the transition to my new lifestyle was enjoyable and not too much of a shock to the system. We had to work hard and the hours were long, but the challenge was a satisfying one – we were, after all, breaking new ground by introducing an altogether novel technology – and I like to think we were successful, despite the almost hourly crises and traumas. Among the aircraft in which we were exclusively involved included the Dove and Devon and later the Comet being built by the de Haviland Company, a generation of new designs from the Airspeed Aircraft Company, and the Canadair DC4 fleet bought from Canada and being operated by BOAC over its long-haul routes.

In due course, in the interests of economy, most of the Division was brought under one roof on the fourth floor of Marconi House, and the free and easy life we had enjoyed away from Head Office while at Writtle largely disappeared. However, it soon became apparent that New Street's production capacity was insufficient to cope with the growing demands of both the Communication and Aeronautical Divisions, and so, in 1954, the latter was given exclusive use of the Company's new factory at Basildon, with some of the staff, including myself, actually moving home into the New Town.

In 1963 I was prevailed upon by my brother-in-law to join the family entertainment company in London and so, with some misgivings, I left Marconi's for The Big Smoke. The two-week probation with Aeronautical Division had stretched into an unbroken seventeen years! Sadly I was not impressed either with the type of business in which I was then involved or with living in the City and, after giving both a fair crack of the whip, I approached Marconi's on the off-chance they could find me a suitable slot.

I was interviewed by Communications, Radar and Closed Circuit Television and opted for Radar, joining them at Crompton's in Writtle Road. I had as a consequence swapped Aeronautical's little sets which were built in weeks and weighed just a few pounds for Radar's massive equipment which took many months to build and was weighed in tons!

The Writtle Road site of some twenty-six acres was strictly Victorian and shortly after my return the Company embarked on an unbelievably expensive programme of modernisation in which I, as Establishments Manager, was heavily involved. This was probably one of the most interesting and exciting phases of my time with Marconi's which only ended when I was appointed Administration and Security Manager at UKADGE in London, an equally exciting consortium of three great companies, Marconi, Plessey and Hughes of America, involved in the development and installation of a unique Command and Control system for the defence of the United Kingdom. That multi-million pound contract was still very much in progress when my sell-by date came up and I retired in 1987.

Without shadow of doubt I can say I enjoyed every moment of my years with the Company and am grateful for the privilege of working with so many real friends.'

Roderick Mackley

More Reminiscences

'I passed out as an RAF Apprentice in 1943 and subsequently had many strange connections with Marconi. My first posting was to RAF West Drayton and found myself doing acceptance testing on SWB8 vehicle mounted transmitters. I also tested the first 1 kW VHF transmitters – I cannot remember who designed them but I hope it was not Bill Barbone, as the designer failed to interlock the door to the lecher bars necessitating a switch off – open door – tune- close door – switch on sequence being repeated many times! In the middle of the night, working alone, my sequencing failed me and I caught hold of the live bar. I was given a knock out pill and woke up the following evening! Since 18 year olds were deemed to be infants in those days, the formal enquiries took some time.

About two years later a group of Cranwell apprentices plus a stray Flight Lieutenant were gathered together on a secret course in a tatty but hidden at the back of the balloon sheds at Cardington. The subject single sideband transmission! The only person who failed the course and was sent packing was the Flight Lieutenant. After many supplementary bits of training at various places including the GPO Cambridge we were shunted off to the sites chosen across the world – in my case to the Egyptian desert.

If my memory serves me right the transmitters were SWB10's. However we had about 60 transmitters on site so that SSB was only part of the scene. Many were American lend-lease and the main distinction was that each sandstorm caused the latter to go into parasitics whereas the SWB8's and SWB10's did not mind in the least.

Here we learned a few things that the natives could do which even the transmitter group could not have anticipated! The most spectacular was to steal the beautiful Marconi timber boxes which contained the termination loads for multi-rhombic antennas in the middle of night shift without electrocuting themselves or tripping the transmitters. The other trick was to cut the best quality cables, hitch a camel and drag a suitable length out. When I finished 14 years in the RAF, I applied to the Company and within a week I was involved in SSB drive and receiver testing!

Subsequent work led to being attached to Pat Keller's group at Writtle to study the testing of the first Automatic Error Correction equipment. In fact I ended up doing much of the design of the seven unit monitor before someone found out.

After that came Widford Hall and the Central Test Equipment Research Unit and a number of years involved in developing automatic testing equipment, etc. I still have the ancient key to Widford Hall! Originally when we took over Widford Hall it was moated. The owner then had the moat filled in and when we went to work on the following day the moorhen which had occupied the moat was sitting in a puddle in front of the house.

The breadth of equipment covered by old Marconi's and the range of expertise of the people involved was really quite staggering, particularly as loyalty rather than salary was the key factor in preserving the know-how! There was always someone who knew the answer.

Post retirement, I enjoyed about 8 years as a visiting lecturer at the Marconi College training about 600 RAF and industry personnel in RF radiation safety and published two books on the subject. During this time we had a spare radar at Bushey to play with! In addition we had mobile transmitters on the College lawn. On the demise of the Marconi College I set up the course at another company in association with Mike Spalding and he has now taken it over and I am, at the age of 79, limited to gardening!'

Ronald Kitchen BEM
[It wasn't Bill Barbone]

'**Although** I was never employed by Marconi's (I worked for a competitor - Plessey) I was a 'customer' in that I was in the RAF when Marconi was supplying the equipment for the Rotor programme of the early 1950s.

With great haste, the Cherry report of 1949 recommended an urgent overhaul and improvement of the UK's air defences, under the codename Rotor. Due to this, it was recommended that the sprawling network of some 170 radar sites left over from the last war be rationalised and consolidated to 66 sites, and that the best existing radar be completely re-built to higher peacetime standards. The essential elements of the wartime Control and Reporting structure were maintained - a hierarchical command and control system, separate sectors etc.

The contract was given to the Marconi Wireless and Telegraph Company and was (and still is) the largest government contract awarded to a single UK firm.

The project was massive. The re-manufactured radar equipment consumed valuable manpower and resources. It must be remembered that the country was under dire economic circumstances at this time, with rationing still in place for many items. But this effort resulted in massive improvements in reliability and maintainability, as well as performance: some equipment had its range more than doubled.

Ironically much of the reporting system and the equipment was obsolescent by 1953 (i.e. within about 3 years) I hope you will be able to mention Rotor in your book as it transformed the Marconi Wireless Telegraph Company'

Norman Bartlett

'I worked at Marconi's 13 years. I started in the Cable Form Department on piece work. There were a pair of twins who could make them so quickly no one could beat their time. I went into Sign Writing. I was in my teens. I was based in a workshop with about 60 men, who were the salt of the earth and very protective of us girls about 2 or 3 of us.

I was married while in this Department and felt mean as I never took any bags or spare clothing in that last day before my wedding leave, as tricks were often played on people, things hanging on the rafters etc. my paints in a tin were screwed down onto the bench once! I was escorted to the Co-op in town by the Shop Steward to choose a wedding present, I chose an ironing board and a mangle, the ironing board is still in use 52 years later, quite a heavy one.

I moved onto Microfilming in another part across the road from the main building. I ended up in an office in the main building where I did filing, etc.

They were a good firm to work for. I often did running repairs when in the Sign Writing Department, sewing buttons on, one chap bent over as I repaired his overalls, the others all started hammering on the benches to draw attention to us, I blushed a bit being so young.

I also had a preview of colour TV, when I was shown one on my journey round the workshops. We all cycled to Marconi's daily. They were happy memories.'

Mrs Lilian Gowers nee Baldwin

'I was with Marconi Wireless Telegraph Co. Ltd for 38 years from 1946 to 1984 at the New Street Factory. My first five years I was working in Receiver Test – testing radio receivers such as the CR100 – CR150 Series, then my chief, Mr J.P. Wykes moved me into the then comparatively new section TV Test where I spent the next nine years. On reflection, I think these years were my happiest.

The Test Department was full of men of great character. J.P. Wykes, the chief, was a character himself. He knew every man in the Test Department by name. On Friday afternoons he would visit one of the Test Sections which were spread all over the Works and he would pick on one particular technician. He would stand behind the unfortunate person for maybe 5 or 10 minutes and not say a word. Then suddenly he would tap the shoulder and ask him if he knew what he was doing, and deliberately get him confused. He would then call the section leader over and tell 'This man does know what he is talking about'! He did it to me once and I was determined he would not do it again. But several weeks later, he did.
Tap on shoulder – 'Do you know what you are doing?'.

Answer, Yes Mr. Wykes, I know what I am doing – Do you? His face burst into a huge smile and he said 'That's the answer I want. I will never bother you again.' He patted me on the shoulder, and left. I still think he was the best Chief of Test we ever had. After leaving TV Test I went on to the admin side of Test until I retired.'

E.H Palmer

'I was born in Portsmouth in 1934 and moved to Chelmsford in 1960 to take up employment with Marconi Radar at Church Green, Broomfield as a Trainee Draughtsman. I then worked at Great Baddow for a short time before transferring to Marconi Communications at New Street in Chelmsford. My office was called the Installation Design Office and my Section designed Outside Broadcast Vehicles. I remember hundreds of cyclists pedalling down New Street at 5 o'clock along with worker's from Hoffmann's. Also, buses were lined up, waiting to transport workers to various outlying districts. If you wanted to eat in the canteen, you had to buy meal tickets at the beginning of the week. Also, you had to provide your own knife, fork and spoon and there was a tank of lukewarm water provided for you to wash them afterwards.

Every Christmas, Marconi's would put on a free party for employees children at the Marconi Club in Beehive Lane with tea, entertainment and a gift for every child. When they extended the Marconi Club, they had Roy Castle to entertain for the evening, and as I was Captain of our Snooker Team, I was invited to go to the show. I also ran our Table Tennis Team and even managed to win a medal, which was presented by the then M.D., Tom Mayer.'

Len Wilkinson

'Anyway, that building was used for the annual Dinner and Dance. *'The Blue Ramblers'* were always the band for that night – most of them worked for Marconi's anyway, but they were great!

On the corner of Marconi Road, right next door to the 'rail track' and the huge iron gates at the side of the main building, was the 'Caretakers' house. I think the Gatemen went home at 9.00pm, when all overtime finished and he was left in sole charge of the whole site! What would 'Health and Safety' think of that in this day and age?

If you walked up Marconi Road, there was a mesh fence on the left, about 6ft high. Behind the fence was what looked like, a fish pond. Mind you, it was man-made, about 18ft square. The water was a bit murky and it did have weeds growing all the way around it. But I never did see any fish in it. My Father said that it was something to do with all the waste chemicals that came out of *'the factory'*. In the summertime, it smelled like pennies soaked in vinegar, acrid.

I left New Street in 1956, went to Pottery Lane, Broomfield and went from being a Tracer to a Draughtsman. My husband was Geoff Edwards. Geoff was born in Gt. Baddow in 1926. He joined Marconi's in 1941 as a mechanical apprentice, having left what was known as The Technical College in Chelmsford.

In 1944 he was called up. He was the first Marconi apprentice to be called up. He spent 3 years in India with the 'Royal Signals' and then returned to finish his apprenticeship at New Street. He spent the rest of his apprenticeship time in the main Drawing Office in Building 46. The Chief Draughtsman was Reggie Mead. He had to spend a lot of his time, with the men – in the main workshop, half-way down 'that long road'.

He too, liked the front building. We all thought it was very posh'. And if you had to report to someone (even a minor person) in that building, you thought you were privileged and made sure you really looked your best. Those were the days!

Apparently, there was one place in that building that was called the *'Inner Sanctum'*. It was the office of 'him in charge' of everything. I think his name was John [F.N] Sutherland and his Deputy was a Mr. R. Telford. It had a fully carpeted room, huge desk and lots of leather chairs. In the middle of the ceiling was a huge cluster of lamps with cut glass shades. Well that's what I was told, but there again, that was only hearsay!'

Mrs Dilys Edwards

'As far as I can remember this photo was taken in the mid 1950s, it was the first of my many placing's within the Marconi Company. We were all filing clerks in what was called machine accounts, can you spot me? I can still remember most of the girl's names and I still get a Christmas card from the girl standing by the filing cabinet.'

Margaret Hudgell

CHAPTER THIRTEEN

New Ideas

For a time the Marconi group of Companies led the world in computer design and development.

In 1951 Elliott Brothers, a British instrument manufacturer established a computer division, later to be called 'Elliott Automation'. Very soon thereafter the Company felt the need to set up its own semiconductor division and in 1966 set up a facility in Glenrothes (the area later to be called 'Silicon Glen'), Scotland to manufacture RTL and DTL integrated circuits. In 1967 their Research group set up an MOS (Metal Oxide Semiconductor) facility in Glenrothes and by 1968 had produced 8 bit computer chips based on MOS technology a full three years before Intel. Incredibly the Glenrothes operation was shut down in 1969 and in a logic defying act, Lord Weinstock completely closed down Elliott's semiconductor activities just as the Integrated Circuit business worldwide was about to explode.

As part of this brief lead in computing technology The Marconi Myriad was an early computer designed by Marconi's in the 1960s and assembled with the English Electric Computer System 4/30 at New Street.

The Marconi Myriad was a 24-bit machine largely built using integrated circuits from Ferranti. These were packaged in small 'TO8' type cans. The architecture was developed largely by the in-house Marconi team that designed similar, but physically larger computers based on SB345 discrete surface-barrier transistors. These machines were used successfully by the Royal Radar Establishment (RRE) in the UK and by the Swedish Government in their 'Fur Hat' defence system. They also provided flight data for UK military Air Traffic Control (ATC) for 15 years. In Australia, Myriads were used as part of the AF/TPS-802 'HUBCAP' air-defence system from 1967-97. The Computers were also used on Cyprus and possibly Malta for ATC.

The Myriad 1 computer was mounted in a small desk format, and was far smaller than any comparable machine at the time. 8-bit paper tape was standard input (the software could handle data input in either the ASCII or the rather idiosyncratic KDF9 character codes) - but a high speed 1000-characters/second (electrostatic) reader (made by Facit) was capable of projecting paper tape across a room in spectacular fashion. A high-speed printer was provided. The major machine cycle time was around 800 nanoseconds, with inner cycles around 200 nanoseconds.

Most early programming was performed in very amenable and complete assembly code. Some use was also made of a subset of Coral 66 known as Mini-Coral. The 24-bit architecture provided a logical and flexible address/data environment but the 15-bit address limited the memory size to 32K 24-bit words. The operating system allowed multiple programs to run concurrently but most systems were coded 'on the bare metal'. Addressing allowed easy integration of external computing and display equipment.

In 1964, a Myriad was displayed at a major computer show in London. To catch the public's attention, it was decided to deploy a model HO railroad layout containing numbered (1 - 10) rolling stock. The public were invited to enter the order in which they wanted to see the train assembled. Immediately Myriad developed a strategy for shunting trucks around the tracks to assemble the train correctly.

Despite the fact that Marconi's Myriad out-performed most if not all US machines at that time the Company decided to discontinue funding and the project closed. By doing so the Company undoubtedly squandered its lead in another technology destined to change the world.

Another potential world beating computer was the Marconi TAC or Transistorised Automatic Computer. But in the end the Company developed only five TAC machines starting in 1959. In 1965, English Electric commissioned Marconi to build a computer that could handle the monitoring and alarm system at Wylfa Power Station. One of the TAC machines ran in Wyfla Nuclear power station on

Anglesey, UK, from 1966 for 38 years until 2004, (from 1968 non-stop), which could make it the longest running computer in the world. It was used to monitor the Nuclear Reactor. Work had begun initially in 1965 to assemble the system in Marconi's New Street Works and then was moved in May 1966 to English Electric's Head Office in Kidsgrove for completion.

Memories of Marconi's at New Street 1962

'I first came to work for Marconi's in September 1962 as a Graduate Trainee, fresh from obtaining my physics degree. I eventually retired from the Company in September 2005, although by that time the Marconi name was a distant memory and the Radar Company I worked for had long since changed into a Division of BAE Systems. The only time I worked at New Street was for 5 weeks at Xmas 1962, when as part of my graduate training I spent some time in the Test Department of the Avionics Division.

The main product in the Test Department at that time was computers for the airborne Doppler radar Navigation Systems for the French Air Force. These computers were state of the art at the time, although any one from the 21st Century would have had great difficulty in recognising them as such. They consisted of a complex mechanical gearbox with electric motors driving shafts according to the output from a Doppler radar sensor with potentiometers measuring how much the shafts had rotated. All this effort drove a set of mechanical dials that showed how many miles in the desired and unwanted directions the aircraft had moved.

However my main memory of New Street at the time was the rigid social division into staff and workers. At lunch time on my first day at work, I decided to go to the canteen for lunch, following my practice at my previous period at Marconi College. When I got to the canteen, I found that there appeared to be a choice of self service or waitress service. I decided that self service was good enough for me and would probably be quicker. I joined the queue, picked up a tray and collected and paid for my food with a previously purchased ticket. I then looked for the cutlery as one normally does in self service restaurant, but was unable to find any. When I enquired, I was told that employees were expected to provide their own cutlery in the workers Canteen. Presumably the workers were not trusted not to steal it. When I was asked who I was and explained I was a new graduate trainee, it was politely pointed out that all graduates were 'Staff' and expected to use the 'Staff Canteen', which was the waitress service area. I did not make that mistake again.

I did not have lunch at New Street again until the late 1990s when I was working for Marconi Radar in the new Eastwood House building at the rear of the old New Street site. Radar employees were allowed to go to the Communications Company canteen. This was by then in a different location, it was self service for all and they trusted the workers not to steal the cutlery they provided.'

Keith Ronaldson

'I first started working for Marconi's in September of 1969. My wife started there the same time as myself. For 22 years it was like working in a large family group. Some you got on with, others you didn't. But through it all we soldiered on. We were in the Printed Circuit Board Manufacturing Process Plant. We made all sorts of boards, both rigid and flexi rigids. Some boards were even made there that fitted in the front of satellite dishes to pick up and decode the signal from the dish itself.

Then came the trouble when Arnold Weinstock was due to retire. After all the dust settled, the Firm's Directors decided in their infinite wisdom to leave traditional market places that had served them and their customers well, and instead to concentrate on the lucrative defence market. So well-known brands like Hotpoint and Schreiber were simply sacrificed and dumped. The in-car mobile phones which were installed at the Waterhouse Lane site; well. they went as well. Everything that the Company had profited from before went.

I am reminded daily of the old Marconi, New St, factory where I started my working life back in the 1960's. It saddens me to see the now derelict buildings looking so forlorn, buildings which were once 'alive' with people working and producing all manner of what was then 'high tech' electronics for the broadcast industry together with communications and radar equipment, a real hive of activity – now nothing, a ghost factory.

As I walk up Marconi Road to my current place of work, I try to picture the now demolished high power test building where I once worked, and as I look from my office window at the Essex Heart radio station building, I see the new BAE building and car park where Marconi Building 46 once stood.

It was September 1968, and having just left school with an interest in radio and electronics, there really was only one place for this local lad to seek employment – The Marconi Company. I was in a way following the family tradition, my grandfather Arthur George Swain had moved from London to become an engraver for the Marconi Wireless Telegraph Company in 1912. He then settled in Chelmsford bringing up a family, which resulted in my father Fred Swain also joining Marconi's before the 2nd world war, as an apprentice tool maker. Following national service with the RAF he returned to Marconi and re-trained as a draughtsman. Now it was my turn to join the company, and I was to become an apprentice technician.

I probably did not fully appreciate at the time, what brilliant opportunity this was, there seems to be little like this available to young people today. In the first year I spent my time split between learning all the practical manufacturing skills at the New Street training centre, many will remember 'the pit', and block release to collage to learn the theory. In subsequent years, I worked in various test departments around the Company whilst continuing my college training. As an apprentice I remember working in Radar test, Marine communications and Test equipment repair and calibration, both in the main workshop by the main gates, and the hut, that still stands as I write in 2012, backing on to Marconi Road. This hut was air conditioned and was used for calibration of sensitive instruments, my job while working there was repair and calibration of Avo meters. During this time I also remember the occasional job taking me out of the factory such as visits to Rivenhall air field where Marconi were also based in addition to the numerous units around

Chelmsford. At Rivenhall I remember being involved in testing waveguides, and on another occasion circuit checking a very large short wave broadcast transmitter.

Having passed all my technical exams, I was now a fully fledged 'Test Technician' and continued to work in various departments in the New Street factory. I mainly recall my time in 'High Power Test', this was the early 1970's around 72 to 73 and I recall working in a small back room that backed onto Marconi Road, here we tested transmitter drive units that were destined for the new local radio stations around the country. I also became increasingly involved in the TV transmitter side of things in the main area of the department. This was a time of massive expansion of the UHF 625 line colour TV network, and Marconi was supplying most of the transmitters. I started off circuit checking transmitter cabinets, and testing transmitter changeover units, before getting involved in testing of the various transmitter sub units, and assisting with the final testing of the complete transmitters. So many transmitters were being built at the time that I seem to remember them being built and tested all over the place, I can definitely recall working overtime one Saturday and helping with the test of a TV transmitter in building 46 at the top of the yard. The lines of transmitters in the high power test section all had the names of the transmitter sites to which they were destined attached, I remember weird and wonderful names such as, Moel-y-Parc, Pontop Pike and Bluebell Hill and used to wonder where these places were.

By the mid 1970's sales of Marconi Mk8 TV cameras were booming and TV test were struggling with the work load. As a result I found myself seconded to the TV test section. This was located in the main body of the New St. building; I enjoyed my time there which even included a period of night work, because we were just so busy. It was a little eerie working in an otherwise empty factory through the night and a strange experience leaving to go home to sleep, just as everyone was coming in to work in the morning.

The boom times in TV test did not last, and by 1976 I found myself away from the New Street works at the Widford site for my last year with the Marconi Company.

At this time an opportunity arose that I could not refuse, at the beginning of 1977 I set off on an adventure on the North Sea. So over the next almost three years I spent my time either on or off the then rather rusty old radio ship Mi Amigo, home to Radio Caroline. Who would have believed that my years of hard work would lead to an opportunity such as this, I was in my element maintaining a complete radio station, and getting to broadcast as well!

After Radio Caroline I went to the BBC, where what I had learnt about TV engineering at Marconi came in very handy as I worked in the VT department at TV Centre at White City. Later in 1982 I moved back into radio joining the London commercial station Capital Radio as an engineer. I stayed with Capital through all the many changes over the years, moving building from Euston to Leicester Square, mergers, and changes in ownership. As Global Radio, which owns many stations around the country, I found myself moving back to work in Chelmsford in 2009, with Essex FM which has more recently been re-branded as Heart Essex.

So now in 2012, I have quite literally come full circle, since starting work at Marconi, New Street in 1968. Looking out from the Heart Essex Broadcast Centre in Glebe Road, I am constantly reminded of where it all began, and how appropriate for a radio station to be located here, as it over looks the old Marconi site, a place with so much real 'radio' history. I hope that whatever happens with the old Marconi factory that the developers preserve as much as possible of the old buildings, as a reminder for future generations as to just how important the Marconi Company was to the people of Chelmsford.

Ed Swain

Opposite is one of the original 2 inch Whitworth bolts used in the masts construction. This bolt has an engraved plate attached, clearly my grandfathers handy work, as a Marconi engraver, with the words :- 'Bolt From - Marconi's Aerial Mast - Erected 1912 - Dismantled 1935'. This would have made my father only 13 at the time the mast's were dismantled, which I think makes it more likely that the photos (of the mast) originated from my grandfather Arthur George Swain.

Ed Swain

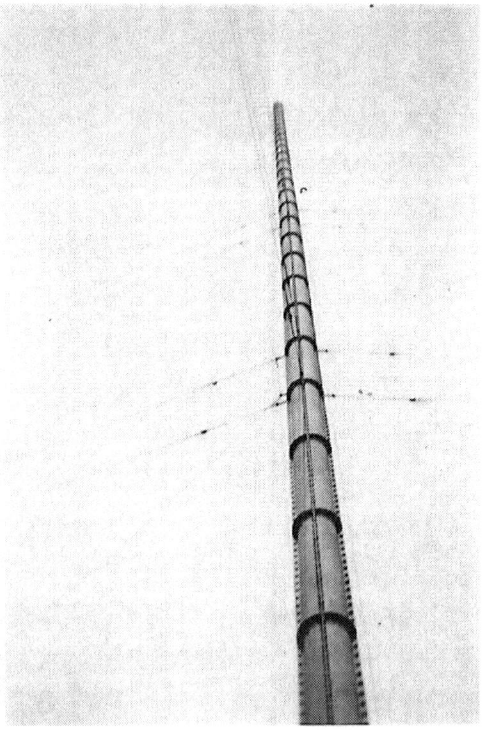

Bolt from
MARCONI'S AERIAL MAST
Erected 1912
Dismantled 1935

Bolt from
MARCONI'S AERIAL MAST
Erected 1912
Dismantled 1935

Employees could also purchase these products direct from the Company at a discount. So you got paid and the Company got some of your salary back when you bought from them. It was all round a win, win situation for all concerned. Plus if you thought a product was really well made and good value, the chances were neighbours would look to purchase that same product through a shop. But all that went, and afterwards the in-fighting at board level didn't help. Just before our Section closed down I was the Vice Chairman of the main Union there, and also a Senior Head Inspector. Telling people that the jobs we had being doing for years were about to end was no fun at all.

The best advice I could ever give anyone was to forget what we had been doing, as it was such a narrow field of expertise. Instead they should look for something entirely different than they had ever done before. What went really wrong was that they were not competitive enough. Now to anyone who drives by, all you can see are the front Main Offices and Security Gatehouse, trouble is that all the windows and doors are boarded up to prevent break-ins.

Even worst news of all was learning that the Development Company who had purchased the site had themselves gone broke. If that wasn't bad enough, the source of their finance had also collapsed, so now it is all in the hands of the Receivers. Even they not so long ago have had notices issued by the local Borough Council to sort the site out, to keep it tidy and to prevent break-ins. Apparently this has already happened. This is such a sorry state of affairs for a once great Company site to finish up in. Will there be any good news for this site; only time will tell? Time and any change in the present financial circumstances we all are facing.'

John Crouch

'I was born to a second generation Chelmsford family. My Grandfather was the Security Guard on the Marconi's gate (Townfield Street end), his name was John Thomas Hanchett, he was the treasurer of the local Salvation Army. My father was Arthur Hanchett, a Musician who played in The Chelmsford Citadel Band from about 1925-1962, he also sang with The Chelmsford Orpheus Choir and appeared at many functions around the area as a Bass soloist. I am proud to have grown up in Chelmsford and have many memories of playing on the old bombsite on Townfield Street where I lived. I also have happy memories of the late 1950s and early 1960s, enjoying being a teenager in The Orpheus Coffee Bar in a basement on London Road just before the river crossing.'

Brian Hanchett

CHAPTER FOURTEEN

New Street Services

The following was taken from *Marconi Companies and their People* the title given to the Company house magazine and was started in the early 1950s. Priced at sixpence and printed on high quality heavyweight art paper, it recorded in words and pictures many of the activities of the Company and its staff. The magazine, which finished around the 1970s was edited in its later years by the indomitable Dusty Miller and his staff. It came out monthly and was eagerly awaited by Marconi employees. For a time in the mid-1980s the Marconi Publicity Department produced *Marconi Newslink* again aimed at informing the workforce of what was happening across the enormous Company.

The Way To Your Hearts

Every working day, over one thousand five hundred people stream through the New Street Canteen. Between them they consume each day over 140lb of fresh meat, 120lb of cabbage, 36lb of peas and 7 cwt of potatoes, 40lb of onions and 18 gallons of soup.

A large variety of puddings complete the meals and for the fifteen hundred there is a daily choice of at least eight, including trifle, queen's pudding, cream gateaux, milk pudding and jam or treacle tarts. All meals are completed within one hour, between ten to twelve and ten to one.

At the centre of this big, daily operation, and firmly in control is Manageress Mrs. Doris Warren. She has carried out this important job for Marconi's since the spring of 1959 nearly six and a half million luncheons provided by Mrs. Warren and her staff have been consumed during these seventeen years, let alone the great number of evening meals for shift workers, and snacks by the thousand.

Her life has been devoted to the tastes of others. At the age of fifteen she started her training in a sea-food restaurant, spending five years learning the many aspects of preparation and cooking involved.

Seven years were then spent in the baking trade, as manageress of a bakery, handling both the baking and sales sides of the business, with five assistants under her control.
From then until joining Marconi's, Doris Warren was Cook Manageress for the Essex County Council, except for a break of eight years whilst her children were small.

Altogether she has spent thirty-seven years in the catering trade. She is no organising figurehead, and asks nothing of her forty-three staff that she cannot do herself. 'In the old days', says Mrs Warren, 'I took a daily hand in the cooking and do so quite often now.'

But these days the pressure is constantly on, and her office could be compared to an Army H.Q. during a military operation. The telephone rarely stops ringing, and people are constantly bursting into the room with problems needing an on-the-spot decision. In addition to feeding the fifteen hundred, New Street buys in bulk for Widford, Church Green, Rivenhall, Writtle and Baddow, and supplies Hackbridge. Cooked meals are supplied daily to the canteens at Beehive Lane and Witham.

The biggest problem today is getting staff. People do not apply for canteen work anymore. 'I suppose', Doris Warren commented, 'it is because we are always working against the clock.' No doubt this is why the staff she has got are of a high standard, loyal, hard-working people, who enjoy the challenge of racing against time. Key figures in the kitchens are Chef Vie Dodd, Pastry Cook Jimmy Adam, Supervisor Lily Willcock, Cook Alison Ewers and Assistant Cooks Rose Bates and Lilian Boardman.

Jimmy Adam every day makes twelve hundred cakes, three hundred bread rolls and eight or more varieties of puddings for the midday meal. He has been with the canteen for thirteen years, having started as an apprentice. His attractive pastries would enhance any table. Up to quite recently, when the Luncheon Club was opened, all the Conferences were done by New Street Canteen, in what were called the Convention Dining Room and the Chairman's Room.

The weather forecast is of the utmost importance to Mrs Warren, she keeps a close check on it and has developed a keen weather-sense herself. This is because of the vital part rain plays in her catering arrangements. If it is raining by eleven o'clock in the morning more food must immediately be prepared for luncheon, as attendance in the canteen will inevitably rise by anything up to one hundred and fifty.

On the other hand, if the weather on a mid-winder's day suddenly turns mild and spring like, canteen attendance will drop dramatically. Caught unawares by a change in the weather, Mrs. Warren could be left with great quantities of food on her hands. But she is never caught napping, that is the secret of her success.

> '**At** New Street they had the Executive Dining Room for Directors and Senior Managers only. With wine and waitress service it was really meant to be for VIP's and visitors only but many of the Senior Managers treied to hide there most lunch times including myself for a short while. I remember Hugh was the Chef and produced a stunning roast beef carvery and even prepared some pheasants for us that we happened to shoot one Sunday. He could also work miracles at very short notice, he once had to prepare lunch for six visiting Indian engineers who were lactate vegetarians who didn't eat eggs or cheese. He ran to the market and produced a platter with every type of vegetable and fruit known to man. I almost disregarded the Beef...but only for a moment. The Luncheon Club was very civilised, you paid monthly directly from your salary, phone calls were frowned upon (no mobiles back then) and the rooms were excellent for meeting customers and colleagues from other Companies. The room also had a sense of history. Then we had a very large new MD called Eric Green take over who decided that he couldn't eat there because of his diet so no one else could and he shut it down.'
>
> *Tim Wander*

'I had two stints at New Street as in 2003 I went back and worked for Selex before they moved to Basildon. At that time the whole of the front block and the main factory was out of use but I remember going for a wander to the old Executive Dining Room and finding it like a time capsule with the menu cards and coffee trays, with sugar lumps etc, still laid out on the tables. Going out the back a fridge full of cans of mixers was gently rusting away. Clearly someone had told the staff to go some years previously and they had just upped and left! To my mind it could have had some future as a 1970s themed restaurant!

Do you remember the Queen's Award for Industry wine glasses that held about 1/4 pint of wine? Ideal for introducing a bit of bonhomie into the afternoon's meeting!

As an even more off topic observation the railway sidings...what I hadn't realised until recently is that that the marshalling yard on the opposite side of New Street was used for other general deliveries other than the mail wagons delivering to the sorting office. One of my neighbours told me she used to go down to New Street with her children to see the circus elephants unloading and walking up New Street to Central Park.'

Steve Readings

Marconi Fire Brigade

The Marconi Fire Brigade had a long and distinguished career using members co-opted from the normal engineering and manufacturing staff. During the Second World War the brigade turned out on every occasion to support the over stretched local brigade during the worst of the German air attacks.

Marconi Fire Engine

The *Marconi* Invincible Fire Engine or Gwynne Light Pump was built by the Gwynne Company of London in 1922 as a demonstrator. It was later sold to the village fire brigade of Mountnessing, Essex and possibly Larma Engineering in Ingatestone. In 1939 it was passed on to the Works Fire Brigade of Marconi at Chelmsford. During a night of heavy bombing during WW2, this little fire engine helped save the machine, paint and carpenters workshops by pumping water at the rate of 350 gallons per minute for no less than 24 hours. The 950cc engine is rated at 8hp and will pump water at 100gpm at 100psi at an engine speed of 3000gpm In 1955 the pump was on display at Essex Fire Brigade HQ at Brentford. 14 years later it was restored by apprentices from Marconi Ltd and is now at Coventry, exhibited in the Coventry Motor Museum.

For a long period it was maintained and demonstrated by members of the Marconi Apprentices Association who drove it from London to Brighton and regularly took it to local events and for Company publicity shows. For a time it had a new PMR mobile radio installed and was used as a backdrop for the Radio Distress Signalling Unit developed at New Street for the Fire and Rescue Services. The fire engine became the bane of the Transport Department as it had to be MOT'ed, serviced and insured that required two Directors signatures to approve. It often broke down and had to be recovered and servicing was not straight forward either, especially as it had a clutch plate that was hand made from leather stretched over a cork base.

> '**Two** of the apprentices names who helped restore the engine were Dave and Simon. I remember seeing this at rallies 80/90s and have many happy memories of this very machine from my time at Marconi's, including driving it from London to Brighton. Can anybody remember any rallies where the back axle broke (crown wheel).'
>
> *Chris Wild*

The photo first appeared in the November 1959 edition of *The Marconi Companies and their People* accompanying a report of the Inter-works Fire Brigades competition for the John Rogers Shield at Waterhouse Lane. In the photo with the John Rogers Shield and other trophies are, back row (left to right): Arthur Leveridge, Percy Holland, Jock Muir, Graham Murdy, Trevor Lodge, Bill Rose, Bert Bartlett, Ninian Pugh. Middle row: R Traylor, Norman Williams, G

McIlveen, Jan Frewer, Peter Buers, Bill Nurse, Alan Munday, Bill Pain. Front row: T Hull, Chief Officer Harry McCarthy, Deputy Chief Officer Bill Munday, Reg Fisher.

Phil Hollington

The Marconi Telephone Exchange

'I joined the Company initially in 1971 for a period of two years, and then came back again in 1973 for a period of one year, then ultimately in 1980 for a period of twenty one years, yes I know, some people just don't know when to give up!! The headcount of telephonists at that time was 23.5, there were three part-timers and each was counted as a half head. The switchboard hereinafter referred to as the 'board' was an 11 position 1A lamp signalling type. You may wonder why there were so many telephonists and only 11 boards, the reason for this was that we were at that time Head Office and thus were responsible for 'manning' the outstations, these were then all over the place:- Writtle Road, Writtle, Witham, Rivenhall, Three Bays, Beehive Lane, Billericay and Basildon.

There was also an inter-company bus service at that time the timetable for which was displayed in the Post Room. This bus service was comprised mostly of 8 /10 seat minibuses and could be used by any employees travelling between sites for whatever reason. In addition to people, all the internal post was transported between sites using this service. Naturally there came a time when these buses became non cost-effective and the service was ceased (but I can't remember exactly when). The board was open from 08:15 until 18:30 each day and 08:30 to 17:30 on Saturday. There was a rota system to ensure that the board was covered at all times and this included lunchtime. At no time was the board closed for any reason whatsoever, even when we had a fire drill someone had to stay back to take the incoming calls and advise customers that there was a fire drill in operation and thus no-one was replying to phones. There's dedication to duty for you!! Incoming calls took priority over internal calls but no-one was left waiting for more than a few seconds before being answered.

Directors' phones were indicated by having a red lamp and these also took priority over other internal calls awaiting answer. The answering speed was monitored all the time and posted each week, if the average went above 5 seconds we knew we were in trouble! It was a stipulation of the job at that time that all operators employed should be GPO trained, this was considered the 'crème de la crème'. Anyone who knows anything about the GPO training of manual telephone operators at that time will know what I mean. It was all about routines and specific phrases and you did not deviate from these. Even now I find I still use many of the same rules when using my telephone whether it be at home or at work. Your position was not vacated without permission from the supervisor, to go to the loo you had to ask for 'a casual', and if someone was already on 'a casual' you had to wait until that person returned. However, you didn't then assume you could just get up and go, you had to wait to be told it was OK to leave your position. It all sounds a bit regimental but those were the rules and they were accepted as such. Miss Gladys Huggins was the Senior Supervisor during my time as a switchboard operator at Marconi, it was she who put the rotas together, she who sent operators out to relieve at other sites and her word was law. I don't think I ever heard Miss Huggins raise her voice, but you certainly knew it if you were in trouble with her. Nobody ever called her Gladys, she was always Miss Huggins, she commanded respect and she got it. I don't think I've come across anyone else quite like Miss Huggins, she was a one-off and I think it's true to say that we all had a very high regard for her. The switchboard used to be located on the first floor behind what was then the ladies loo.

This has all changed now of course, the last I heard was this area was being used as a dark room, I don't know if it still is. The ladies loo has long gone, it was converted into a large office and used for some time as the Registry until the demise of that department some years ago. Each day, the operator on the early duty (there was usually only one at 08:15) had to clean the board. This entailed sweeping all the positions, between the plugs and the keys with a paintbrush kept solely for this purpose. As the plugs were made mostly of brass these became very black and dirty from the natural oils on our hands, sometimes at the end of the day, the side of your thumb and index finger where the plugs were 'flicked' up, were ingrained with black from the brass. A GPO engineer came in periodically to effect repairs etc., and would sit and clean all the plugs on each position with a proprietary brass cleaner. It was important to make sure that the plugs were kept in pristine condition, if they became too dirty or the tips worn then the contact between the plug and the edge of the jack (the hole the plug went into) would be noisy. Thus the operator would be unable to check whether the extension was engaged or not. Inserting a plug into an engaged jack was a cardinal sin, as it cut the call off ! So it was important that all the equipment was kept as clean and free of dirt and dust as possible. Each position had 17 pairs of cords in two parallel rows, one cord for answering the call and the other for connecting to the receiving end. In front of the cords were two rows of switches or keys, these keys corresponded with the cords, the first row of keys were the answering keys and second row the ringing keys. To dial out there was a key located next to the dial and this had to be pushed forward to engage it. It all sounds terribly complex, but it's like everything, once you've learned it it becomes second nature and you do it without even thinking about it. Each pair of cords was coloured:- Red, Yellow, Blue, Green alternately in the same order to the end of the row. You worked the board from left to right and always sat with an answering cord in your hand ready to take a call. On a busy day (Monday usually), it was nothing to have all your cords connected, and if this happened you sat with your arms folded so that the supervisor could see that you had a full board and were unable to continue. If the position next to you was vacant, you took that over and worked two boards at the same time.

There was no smoking allowed at the board and certainly no eating! Occasionally a packet of polo mints was offered up and down the board, but nothing larger than that or you couldn't speak properly and we all know what it's like being on the end of the phone when someone's eating or crunching a sweet! It just wasn't acceptable. We each had a pad on which we recorded outgoing long distance and overseas calls, the completed sheets were collected at the end of the day and a record kept of these sheets for accounting purposes. All overseas calls were sent down to the 'international' position as these could be time consuming and in order that they could be dealt with as swiftly as possible, one operator was assigned to 'international'. It has to be remembered here, that no-one had any kind of dialling out facility of their own, everything came through the board.

We take overseas dialling for granted now, but it's not so long ago that all these calls had to be booked through the International Operator at the local exchange. A 'slot' for the call was then booked via this operator and we on the switchboard were called back when the call was successful (or not). Marketing people were the biggest users of this facility, calls to China and UAE were the most difficult to get connections to, it was not unusual to spend days trying to get a connection. Calls of this nature which took a great deal of time and effort to connect were strictly monitored to make sure that once connected they were not 'inadvertently' disconnected by our end. One aspect of this type of call was that it could take days to connect and then only be in progress for a few minutes. It was frustrating, but the cost of calls then was quite prohibitively high in comparison to call costs today. No two days were alike in the life of a telephonist back then, as the job was totally manual there was this also 'personal' element to it. You got to know people's voices and associated them with their extension number. Some voices were far more distinctive than others and you only needed to hear them a couple of times to recognise them, many users were quite 'chuffed' when you recognised their voice and could refer to them by their name instead of having to ask 'who's calling please?' Some of the Directors became almost offended if you had to ask their name, they expected you should know who they were without having to ask, they never took into account the fact that you may be new to the job, that was not their concern.

The M.D. when I joined in 1980 was Sir Robert Telford, there was nothing 'uppity' about him, if his Secretary (Pauline Easton) was out of the office or had left for the day, he was quite happy to make his own calls and never shouted if he was kept waiting. At the end of the day between 17:30 and 18:00 there was usually only one operator left on the board. If Sir Robert picked up his phone (Ext 192), and you were already dealing with a call you couldn't just abandon that call, you saw it through. If Sir Robert had put his phone down by this time, you'd ring him and ask if he still wanted to be connected somewhere, he was always most appreciative of little services like this and never took it for granted.

There were of course exceptions, and these people could make your life a misery if you let them. There were the impatient ones who if they were not answered immediately would start 'flashing' - this was what happened if the telephone receiver was 'jiggled' up and down, it made the lamp flash. This was a signal for an emergency call and we would always answer with 'emergency which service please?' Naturally, there were some who abused this and thought they could get away with it every time, but there was an unspoken rule for these people, 'make them wait and give them something to complain about'. We didn't have to do this very often, but when we did, the abuser of the system was made fully aware that we all knew what he/she was up to. Fortunately there were not too many incidences of this kind, which I like to think says a lot about the telephone service provided at that time. Saturday working was part of our contractual hours but nobody particularly liked working on Saturday because it was so quiet, as with all jobs, when it's busy the time went that much quicker, but Saturdays were a real drag!!

As with everything, time moved on and we eventually had to replace our beloved switchboard with a modern version. So in July of 1983 all the operators were sent to Wellingborough on a two-day training course to learn to operate a new electronic switchboard. Our numbers had been reduced quite dramatically by this time, many of the outstations had gone so the days of relieving for sickness and holidays etc., had long passed. The new electronic switchboard required only six operators plus one supervisor for the administration work. This was a time of great change and learning for the whole company, everyone had to get used to using this new switchboard. Directors were initially the only people allocated DDI lines (Direct Dial In), but these gradually increased until it's now the norm as opposed to the exception to have a DDI line. Initially too there were many users who could only call internal extensions, or perhaps only dial local calls. Many modifications have been made to the software of the new exchange over the years to accommodate the ever-increasing demands of the users.

As with everything in this throw away society though, this 'new' switchboard is now obsolete (and has been for many years), so the next step will be the scrap heap I guess, but I doubt it's passing will be felt as much as the old 1A lamp signalling board. The personal touch has all but gone with the ability of the users to get their own calls. It was a sad day when that board was finally 'dismembered' and its 'innards' piled in heaps in the yard. Because of where the switch room was situated there was some difficulty in getting all the equipment out, it was eventually achieved by lowering it out of the window above the staircase which leads up to the present post room on the first floor. I can't tell you whether or not any of the positions were put into a Museum or not as I have no idea. I would like to think that this is the case, but I don't even know anyone left here now who would have any memory of this grand old piece of history to be able to tell you.'

Phil Hollington

Early Days of a Secretary at Marconi's

'I can recall when I was a trainee on an Intensive Secretarial Shorthand-Typist Course. I learnt to 'touch-type', using a manual typewriter at the local College of Further Education. We were not allowed to look at the keys, all we had was a diagram of the keyboard at the front of the Classroom. During the training period I also had to type to some military music without looking at the keys! What fun! During the first year I spent 6 months at the Company's Secretarial Training College, the keys of the manual typewriter were covered up completely!

During the first 4 months of my training course in the Company, I worked in a Department learning about systems and office practices and had access to a manual typewriter which comprised of the following: the carriage belonged to an Imperial 66 and the main keyboard to an Imperial 70 with a very large typeface (which was difficult to change the size). One of my first tasks was to type lots of columns of figures which proved to be a nightmare with such a typeface! I felt that the typewriter was fit only for the Museum! Fortunately, I shared the office with another secretary who worked part-time, so I saved up the difficult tasks to use her manual typewriter which was far superior to mine.

At the end of my first year's training, I spent a fortnight in the Technical Information Department where I used a manual typewriter which included technical symbols. I can remember that some symbols I was required to type I had to make up by using a combination of two or even three keys at the same time (so different from modern information technology)!

p.s. In the distant past I have vague recollection of Telex machines and stencilling!'

Val Cleare,
Ex Marconi Comms (now with AMS)

CHAPTER FIFTEEN

Apprentices

From its earliest days Marconi's had a superb reputation as a training organisation. An apprenticeship there was considered to be as good or even better than a University place and the practical experience that could be gained was second to none. Unfortunately by the time I joined, the Company also had the reputation of being great for training, but rather old fashioned to work for.

By 1980 the New Street Works was showing its age. Repairs and maintenance were eventually subcontracted out, but costs climbed every year and cut backs were inevitable. Plumbing was antiquated and the famous long manufacturing area was clearly past its useful life. The main problem was how cold it got in winter and the heat in summer under the glass roof. Often many fans and trays of cold drinks could not dissuade the Union rep from shutting down the line as it was too hot to work.

But in the early 1950s to be a Marconi student at New Street was something special.

> 'I joined Marconi's in 1948 as a technical apprentice. I remember the Goods Wagons shunted across New Street from the Goods Yard, and the traffic had to wait.
>
> The girls Pavilion still lay near railway embankment as the New Street site was the cricket ground. The girls pavilion was demolished and building 720 was built, I worked on the Outside Broadcast vans built on the ground floor, with the canteen above. The apprentices were based in the 720 'pit' but we did our apprentice training at Pottery Lane off Broomfield Road in a Nissen hut.

Remember that Marconi's made everything in its 'in house' Departments including publicity, printing, carpenters, electricians, plating and paint shop, vehicle maintenance and later printed circuit boards. During the war the vaults under the main building were used as shelters and Marconi House roof was used by ARP wardens. '

'Robbie'
ex Marconi House 2nd floor.

'When I was in my final year at school, *'Wireless World'* was my favourite magazine. Every month, in the business section, there were detail listings of the various contracts won by British companies for the design and supply of radio communication systems being installed throughout the world. Marconi Wireless Telegraph (Chelmsford) figured prominently on those lists.

To an impressionable 17 years old, Marconi's at Chelmsford seemed the obvious place to seek employment and training. A letter to the Education Department at New Street Works resulted in an offer of a five-year 'sandwich' student apprenticeship, subject to my satisfactorily passing the Advanced Level Examinations in Pure Maths, Applied Maths and Physics. Needless to say, this was all the encouragement I needed to do as well as I could in that final year at school. I passed the exams, advised the Education Dept. and was duly summoned to turn up at the Brooklands Apprentice Hostel on a particular day (which eludes me) in September 1957. It was also suggested that if I had a bicycle it would be advantageous to bring it with me for getting to and from the factory in New Street.

As my home was in West Wales, my Father decided to drive me to Chelmsford. My brother's 'hand-me-down' bicycle was dismantled so that it fitted in the boot of our trusty Hillman Minx, and with a friend of my Father's for company, we journeyed without incident to Brooklands in New London Road, Chelmsford. Welcomed by the Managers of Brooklands Mr and Mrs Cuthbert, Dad and his friend were provided with tea and biscuits, the bicycle was reassembled and put in the bicycle shed. I had arrived!

Brooklands Hostel comprised two 3-storey houses 'joined together', with the majority of rooms set up as dormitories with up to five students in each room. There were a common eating area and a lounge. The Cuthberts lived on the premises and Brooklands was perfectly suited to its purpose of accommodating around 40 'inmates' who came from all parts of the country, from other countries and from other organisations within the Marconi / English Electric group. In the back garden there was an annex (the Bungalow) which had further dormitories, but I have very little recollection of that building. We were allowed to stay at Brooklands for 12 months, which gave us ample opportunity to 'find our feet' and to make our own accommodation arrangements when it came time to leave.

As you can imagine, the Cuthberts had a difficult job. Accommodating 40 testosterone filled male adolescents had its moments! By and large however, they handled their job well and I have fond memories of them both.

The big attraction to me of Brooklands, however, was the Nissen Hut in the back garden. Set up with benches around the walls and a large black 'pot bellied' stove in the centre. The hut was available for our use as a hobby shed, and during cold weather that stove would glow 'cherry red' when really stoked up! We all came from different backgrounds but we had similar interests... radios, electronics, old motorcycles and cars. There were a number of radio 'hams' in the group and many short wave listeners, these activities being undertaken in the hut. Practical jokes were common place. A kipper wrapped around the output valves of Jimmy James' powerful communications receiver (Marconi CR100?) left a less than pleasant aroma around for weeks. KT66 output bottles sure run hot! No names of course, but we all know who you are Mike. Others in the group have mentioned the 'back gate' giving quick access to the Transport Café on Chelmsford Bypass for a Friday evening snack... usually bacon, eggs and chips! Whilst I do not remember any of that, I certainly remember our Saturday morning trips up Wood Street (just behind Brooklands) to Joe's Scrap Yard. Joe seemed to have some arrangement with Marconi whereby obsolete marine communication equipment finished up in his yard, and Joe (bless him) had no objection to our clambering over all that stuff and stripping out whatever was of interest.

Then came the negotiation over price, and despite their reputation for hard bargaining (I believe Joe was Jewish), we came away thinking we had secured a bargain… Joe probably knew better!

Obviously this was long before 'elfn'safety' regulations came about to protect us from ourselves. Joe's yard was an Aladdin's Cave of 'treasures' to anyone interested in radio equipment.

The nearest pub, and within staggering distance of Brooklands, was 'The Sir Evelyn Wood' on Widford Road. Mike Plant's aperitif of choice in those days was 'Vimto'. It was not long before the Barman taking our order would suggest 'And would it be the usual gin and Vimto for the young gentleman?'. Rumour has it that Mike's drink of choice these days is a little stronger than Vimto.

Directly opposite Brooklands on New London Road was the St John's Hospital Nurses Home. Occasionally we were invited there to a dance. On one occasion, I can recall lifting a toilet seat in the bathroom and seeing the words 'Oh good, there is a man in the house!' written on the underside of the seat. These days we often hear of women complaining about men leaving the toilet seat up… obviously, it has not always been seen as a problem!

Looking over the faces, many memories come flooding back. I'll recall just a few, taken at random: I remember John Dorkins (Dawks) and his inseparable friend John Braddon hailed from the West Country, and soon both acquired a larrikin reputation. The horizontal milling machines in the Training Centre had power feed to the milling table which, through machining necessity, took ages to get from one end to the other. On one occasion near knocking off time, Dawks set the table moving from one extreme of its travel and went off to wash his hands rather than hanging around waiting for the table to centralise.

Through miscalculation, or his just having forgotten (and not noticed the dividing head bolted securely to the centre of the table), the inevitable happened. There was an almighty crash and the cutter started to chew its way through the dividing head. White dust coated Instructors appeared from everywhere! Brad went to tell Dawks what had happened... but neither reappeared for quite a long time!

Trainees, whether Craft, Student, or Graduate all passed through the The Apprentice Training Centre: ATC (Building 720) where training was given in all practical aspects of metalwork. This included skills such as use of hand tools, wiring and soldering, sheet metal work, brazing/welding, material hardening and tempering, metal turning on centre lathes, mills and production manufacture on capstan lathes. I am ashamed to admit that despite all that the Instructors taught us, I can only remember three names... Lane (hand tools and centre lathes), Cooper (capstan lathes), and Whitaker (sheet metal work). There were at least two others... soldering and wiring (who could forget having to 'unsolder' the components from those tag strips after ensuring we had made a really good physical and electrical connection!), and the Instructor on mills.

On the first day I was in the group with Mr Lane who gave me a lump of iron and told to 'file that into a one inch cube son'. Away I went and after much perspiration and checking with my 6' metal rule duly returned the cube to 'sir', who took the 'Moore & Wright' micrometer from his pocket... I was given another lump of iron and first realised that maybe in this place 'good enough' will no longer be good enough!

Over the next three months I gradually acquired lifelong skills, which have since served me well. I particularly enjoyed using the centre lathes. Those Colchester Student machines were excellent examples of their type. The (Herbert?) capstan lathes too were fun to set up, but the subsequent use of them to produce a large number of identical components required elsewhere in the factory could be exceedingly boring! My technique to ease the boredom was to set myself almost impossible targets to be achieved by (say) the next tea break.

The fact that the capstans were near the high side windows of the ATC did not help as I was still a little homesick, and through those windows I could see the trains passing by. I fantasized hopping aboard one for a quick trip to Fenchurch St. Station, whip around the Underground Upper Circle to Paddington, and jumping onto the Pembrokeshire Coast Express... I could be home in no time!

Thanks to the foresight and good humour of those Instructors we came away from the ATC with a collection of tools which we had made ourselves, many of which I still have and are in regular use: The (in)famous Marconi Toolbox... oyster grey hammertone. Also, the yellow duster with which we were encouraged to line the tray and inside of the toolbox. Brass boxes (silver soldered) and screwdrivers 0, 2, 4 and 6 BA (turned, hardened and tempered) Junior Hacksaw (Rod wrapped around pegs on a board) Centre Punch (much battered now) (turned, knurled, hardened and tempered) Each week, out of our apprentice pay (£7/7/6...?) we were encouraged to purchase, or save for, an appropriate hand tool. That Moore & Wright 1' micrometer in the toolbox tray took quite a few weeks! There were a couple of places in Chelmsford which stocked a wide range of quality hand tools... Grippers in the High St. and Vinalls up Moulsham St. Who can forget those big ads on the side of the Eastern National buses... 'You can get it at Grippers!'

Introduction to the New Street Factory:

Brown's Test: Involved the impregnating of transformers and inductors with that effective, but foul smelling brown varnish, and the subsequent testing of the finished items. That smell seemed to saturate my clothes and hang about in my lungs. Fifty three years later however, I can confirm that it does not appear to have done any lasting damage. Brown himself never appeared particularly happy, which I think did influence the morale in the Section. As for when working on capstan lathes, I would set myself impossible targets to be achieved by tea break to relieve the boredom. Some time later, having misconstrued my tactic for a good work ethic, I discovered that Brown had given me a really good report, for which I must be grateful!

Dawson's Test: Involved the incoming inspection and testing of all small components. This would not be required nowadays of course as the Company would only purchase components from Quality Assurance Certified suppliers, so further inspection and testing would not be considered necessary. That is the theory anyway... dream on!

Much of this repetitive work was done by women. One instance when working in Dawson's Test I particularly remember was when he severely disciplined one of the women for eating an orange at her work bench during her tea break. He convinced her that the acid on her hands would subsequently get on the components she would be checking, and which many years later could be responsible for the crashing of an aircraft with the loss of hundreds of lives... and SHE WOULD BE RESPONSIBLE FOR ALL THOSE LOST SOULS! She burst into tears, but the point was made, which I am sure she never forgot... I certainly haven't.

Assembly and Wiring (Shop 17? *[16])*: My credentials in the Wiring Shop were severely tarnished... the tradesman I was working with needed new tyres for his car... his bonus depended on how many of the power supplies we assembled and wired got through Receiver Test without wiring problems. Between us we had shifted a pile that week, only to find that half of them (the ones I had wired) were back on our bench... they couldn't plug the octal valves in. I had used too much solder which had run down through the terminals and filled all the valve pin sockets. They all had to have the sockets removed, trashed and re-wired... needless to say there were no new tyres that week and I was not flavour of the month. I rationalised by claiming if MWT had used better quality sockets the solder would not have been able to run through! So ended the first year of my student apprenticeship... only four more to go!'

Keith Thomas

'**My** interest in electronics 'sparked' when I was around 11 or 12 years old. My Dad bought me a DIY crystal set - probably from one of the national comics such as Hotspur, Lion or Eagle. It really was very basic, not quite as archaic as the cat's whisker variety; but I did have to wind a coil around a cardboard former, solder the connections to the rectifier (small plate variety), a fixed capacitor and to the headphones (ex WWII) type. Then we had to rig an extremely long aerial from the top of our house, to the end of the garden making sure that it was well isolated via ceramic spacers down to my bedroom. Tuning was extremely difficult as the knob merely moved a wiper contact along the coil to get it to resonate at the desired frequency but it worked and I spent many happy hours trying it out and trying to improve it.

The big jump in my interest in electronics came in February 1953 when the south east coast of England, including large areas of Harwich where I lived, was flooded. Hundreds of radios, record players and a few TV's were submerged in seawater and were taken to the local tip and dumped. A good friend of mine, a couple of years older than me and a Corporal in the Air Training Corps brigade of which I was a member, realised that even though the sets were to all intents and purpose useless, there was nothing wrong with the valves in them so we would go down after school each day and take out as many as we could. If you remember the old egg crates, well I had five or six of these filled with valves and my pal had the same.

From then on we were in the radio repair business and I was hooked. We started out just using simple substitution – if the heater didn't glow it was automatically replaced, then as we slowly acquired more experience and more old sets we learned to read circuit diagrams and became expert enough to not only replace valves but resistors and capacitors as well. All this time I was at High School learning about electricity in Physics, operating radios in the Air Training Corps and generally building my knowledge of radios so by the time I was 16 and doing O Levels I knew that my career had to be involved with electronics and I wrote to MWT about apprenticeships. I had a nice reply offering me a position as a Craft Apprentice but also suggesting that instead I think about taking A Levels and joining them at age 18 as a Student Apprentice. Consequently September 1957 saw me standing at a bus stop in Harwich waiting for the Grey Green Coach to whisk me off to my new life while my Mother silently wept as 'her boy' left home for the first time. But being a weekday only Boarder at Brooklands she saw me again the following Friday night so that was all right then.

Compared to some others of the new apprentice intake, I found the first year at Marconi challenging practically, academically and socially. Practically I could solder with the best of them, I was good at wiring, I knew the resistor code and could read a capacitor but metalwork – forget it, I was hopeless at filing, grinding and all the skills required to shape metal into useful tools. I marvelled at the skills that Dorks and Brad had in the Apprentice Training Centre and envied how they could operate the machines with such confidence, but then they did go to a Technical School while I was a Grammar School boy. But like many I still have the toolbox I made with its little yellow duster, my rectangular brass box, my G-Clamp, my Toolmaker's Clamp and my set of screwdrivers hardened and tempered by me from silver steel and with their capstan turned wooden handles and brass ferrules. They are well used now but still perfectly serviceable and have travelled all over the world with me.

The ATC taught me skills that I shall never forget and I still take pride today in being able to make brackets and bend and shape metal. There are just two reminiscences from the ATC that have really stuck with me; one being the number of 2 and 3 feet rulers neatly cropped down by a few inches and a few 64ths having been left in the big sheet metal cutters when the pedal was pushed down, and the other was the *comeuppance* of a particularly obnoxious young Craft Technician working in the stores. This young spotty twit refused to hand over some particular drills needed by another apprentice - he pushed them through the cage and then withdrew them several times teasing the apprentice to grab them. The apprentice got so frustrated that he seized the spotty youth in a headlock, forced his head over and proceeded to squirt oil from a nearby oilcan into his now vertical ear cavity. Cue much squealing, much shouting from the Instructors, but no great harm done and needless to say the Craft Apprentice didn't ever mess around again.

Being a weekday Boarder I didn't get to know Brooklands particularly well and my lasting impressions are of the single bath at our end of the house with a partition wall around it, (over which it was possible to look at whoever was in the bath even if the door was locked) and having marmalade on my toast at morning breakfast. That may sound strange to some but it was the first time I had ever had marmalade in all my 18 years, it was never on the table at home, my parents had sweet tastes and it was always homemade jam.

The Cuthberts who ran Brooklands are a dim memory but one of their staff, Mrs Butcher, later became my landlady for a year or so. We met in Chelmsford, got talking and as I was unhappy with the digs I was in, she offered to take me in. We parted only after she became entangled with a new man in her life who resented having a stranger in his house.
'

Sam Woollard

'It must have started when I was 12 or so when my parents asked me what I thought I wanted to do when I grew up and left school. An Engine Driver came to mind but never materialised. My Father was a radio amateur (G6OH) and I always helped him (he may not have thought so) in his Radio Shack - a somewhat basic shed that allowed peace and quiet from whoever.

 I was at school at the same time as John Cleese (of Fawlty Towers and other fames) and we had three things in common – we were both the same height (tall), we both disliked sport and we both detested dressing up as soldiers in the School Cadet Force and parade around the quadrangle. The Signal Section was the best alternative for both of us although we still had to wear army uniforms. I also had the added interest of the Army Band where I played trombone and a few other instruments in the following years – and probably not very well.

This background had set the scene – being a practical person, as opposed to being an academic, I decided I wanted to work in an electronics company. My Father jumped on this and said I just had to work for Marconi Wireless Telegraph Company because he knew the Managing Director (Mr. Sutherland I think it was but not certain) and he would fix it. Well an interview was duly arranged and I had to travel from Bristol via London to Chelmsford, which was a long way, and that I would be met at the other end. Arriving at Chelmsford Railway Station I went to the Rolls Royce at the entrance and said – Marconi? I forget the name of the guy but he was the one who interviewed me and used the MD's car because, presumably, he was told to fetch 'young Samways' from the station. To me, an eighteen year old (maybe spoilt?), this type of service was quite normal because my father had a chauffeur driven car (but not a Rolls) for many years.

I joined the Company in September 1957 as a Student Apprentice on the second 4-year sandwich course to be run. This entailed six months of each of year in a Marconi Department, the other six months at the Mid-Essex Technical College. The majority of apprentices that started with Marconi, spent their first year at the Brooklands Hostel which was very convenient because many students lived a fair way away. However, I was determined to go home to Ascot at weekends because a) it was cheaper at £2.10s per five days and b) how else would I get the washing done?

The first six months at Marconi's nearly killed me; standing on your feet from 7.30am until 5.30pm when you are not used to it was, and still isn't, funny. Also, we could not work out why us 'electrical' apprentices had to spend six months in the Apprentice Training Centre (ATC) drilling, filing, reaming, milling, welding, painting, folding etc etc when we ostensibly, or more probably hopefully, would become brilliant electronic engineers. I have to say it didn't take long to understand why – married life saw to that!

Some highlights in the ATC, if you can call them that, come to mind:

Each lunchtime the ATC Supervisor used to sleep for 30 minutes on the floor in his office and wake up very refreshed. We knew when all the machinery suddenly went quiet because Jackie Bartlett was walking between the offices. A milling machine suddenly went haywire making a penetrating screeching sound. The culprit was John Price, of course, positioning one of the hardened steel blocks in the way of travel. On one occasion Lord Nelson of Stafford (Chairman of English Electric at the time) came to the ATC and politely but firmly told me I was holding my file incorrectly and proceeded to show me the correct way.

I made a big mistake of lighting my pipe with the oxy-acetylene torch. It disintegrated, singed my eyebrows and went everywhere. I have never touched a pipe since.

During the course of these six months all us apprentices made their own tool boxes, a selection of screwdrivers, a saw and other things I've forgotten, which is probably in the best! Most of these I still have and use.

Then we started our first six months at the Mid-Essex Technical College (METC), and what a change that was – we could sit down! We did physics, chemistry, thermodynamics, electronics, heavy electrical etc in both theory and practical. Over the four years we were whittled down to just eight so the pruning was harsh. I must add that we had a girl called Sheenagh Logan who was one of the eight finishers. She had a photographic memory and spent much of the lecture time knitting in the back row. I wonder where she is now?

During our final year we had to implement a project. In our group was John Price, Roger Smith, Fred Hoyle, someone I can't remember and myself. We invented SECS – Student Electronic Computer System. I remember it cost the METC a lot in discreet components and I think it roughly worked but with questionable accuracy. It used the (a + Jb) notation if that still means anything to anyone.

I have very little recollection of which Marconi Departments I went to but they certainly included Great Baddow, Writtle and New Street. Yes, the coil shop at New Street comes to mind. Our bosses used to send us apprentices – one at a time – to walk through the Coil Shop. The chitter-chatter of the 200 women employed there suddenly stopped. How embarrassing! We learnt very quickly!

And so we came to the end of our four year course and it was time to go into the outside world and get a real job. We were fully trained electronic engineers, boat builders and general dogsbodies and my first job was installation work for the Marconi USAF contract and; wait for it – mixing concrete!

I spent many happy years with the Company and after stretches with English Electric Leo Marconi at Kidsgrove and London I left in 1968 to join Honeywell Computers.'

David Samways

'I had enough trouble trying to get to New Street and back.... and that reminds me of another story for the only cycle I could get hold of belonged to my eldest sister Rose, so consequently it was the only ladies bike in the Brooklands shed. The bike had nowhere to carry books to and from METC so one night (and I think Mike Plant was with me) as we walked up one of Chelmsford's side streets, we spotted a green Walls Ice Cream bin outside a local shop. As always I had a small screwdriver with me (try carrying one of those nowadays and see what the police say) so with Mike watching out, I loosened the screws and lifted the bin off the wall. A couple of brackets made up in the Brooklands garage and from then on I had the cheapest book carrier in the town – and I never painted over the Wall's logo...

I have many recollections of John Price – first the swagger, the bravado, the confidence that this young man had; my immediate perception was that this guy knew his way around – he impressed me in the same way that Del Boy in Only Fools and Horses impresses you. So much *chutzpah*. Secondly I am sure that he told me he came from Royston – correct me if I am wrong John, and I had never heard of the place but thought it was a suburb of London.

John started his first few days in the Apprentice Training Centre in a brand new boiler suit, only changing into the drab grey lab coats when forced to conform to MWTCo rules. And what's more I think he was a better ballroom dancer than me, at least I seem to remember him being Mr Twinkletoes when we stepped out with the girls at the dances organised by the Cuthberts!

Academically I found the last year of my Sixth Form education at High School hard going in Pure Maths - Calculus was particularly challenging and I only just scraped through for my A Level pass - so when we did get out of the Marconi Apprentice Centre and into Mid Essex Technical College, I struggled there too. In fact I struggled so hard to master the technical course syllabus that I neglected possibly the easier parts, and when it came to the final exams I failed one of the social requirements – either Industrial History or Industrial Law and wasn't there a Social Psychology of Industry?

I was bumped from being a Student Apprenticeship down to Technician Apprentice and spent the next three years either in 'the shops' or in Receiver Test and had to pursue my academic studies at Night School. There were no 're-sits' in those early days of Higher National Diploma, so it was ONC and Higher National Certificate for me from then on. I revelled in the new system and flashed through ONC and HNC in a further 3 years, got out of my apprenticeship a year early, continued studying for another 2 years while working at the Research Labs at Great Baddow finally emerging with the Diploma in Electrical Engineering issued by the IEE. I was done; I was the electronics engineer I had started out to be when I built that first crystal set all those years ago and I look back on my time with Marconi with great affection and a huge amount of gratitude. It was the best training that any engineer could wish for and I am proud of every moment that I spent there.'

Sam Woollard

'I started working for Marconi's at New Street Chelmsford in early

September 1969. At that time, all apprentices arrived at the Works on the same day, and induction started in one fell swoop. There were four main types of apprenticeship – student (those destined to read engineering at University), technician (those who would be studying for the ONC/HNC at the local College), and craft (those destined to follow the City and Guilds route), and commercial (the few girls who would be studying at the local College for clerical and commercial certificates). We had different patterns of study – the students went to University during the normal terms and then came back into the factory during the vacations, the technicians attended College for half of each academic term and were in the factory for the rest of the year, and the craft and commercial apprentices were basically factory-based, and had day release one day a week during term-time.

On that first day, we all congregated in the open canteen at the top of Building 720, and, as someone who had attended an all-girl Grammar School, it was a heck of a shock to arrive and see ranks of pimply adolescent boys – the craft and technicians were mostly 16, while the smaller number of students were mostly 18. There were about 200 of us altogether, of whom I remember 5 being girl commercial apprentices and I was the one and only female engineering apprentice – something I had not expected. The first morning, we were greeted and briefed by someone from higher management, who told us what a wonderful choice we had made, and I remember filling in numerous bits of paper, then we were shown how to queue up for lunch, using tickets that we purchased for our meals, and in the afternoon, we were broken up into our smaller groups and a more detailed induction took place. It gradually dawned on me that I was going to be on my own, the only female engineering apprentice, for the next four years. Had my boyfriend not been living in Chelmsford as well, I think I would have turned tail and run.

I am told I was the first female technician apprentice the Company ever had. I never set out to be a pioneer; coming from a family which had no gender stereotypes it had not, naively, occurred to me that girls did not become engineers. I was curious for many years about why Marconi's took the 'risk' of employing me, as six other companies that ran Engineering Industry Training Board apprenticeship schemes had turned me down flat. I was later told that they were so surprised when they received my application that they decided to interview me, and when they did so, thought I had the grit and determination to succeed. I certainly really enjoyed the apprenticeship, although it was tough; from being in the top quarter of girls at school (but not the top), I found that I achieved top marks relatively easily, and at the end of my apprenticeship was Technician Apprentice of the Year. I also won the Chelmsford Engineering Society's Apprentice of the Year award in 1973, and became a speaker for the Company, promoting the idea of attracting more girls into engineering, a campaign of which the Company was a strong supporter. Most girls recruited from school became student apprentices, although a few followed me into the technician route; by the time Women in Science and Engineering (WISE) Year was held in 1984, around 10% of Marconi Communications Systems Ltd engineers were women.

I encountered no real discrimination until the management regime changed, and very little prejudice. The only barrier I encountered was very early in my career, when I found I was not being considered for further training because of stereotyping. If I used the title 'Mrs', it was no use training me as I would be leaving to have a baby; if I used 'Miss', I would be leaving to get married, and if I used 'Ms', I was, horror of horrors, a militant feminist. I took the simple step of abandoning all titles, and to this day, my professional name is Barbara M Stephens. From then on, I received equal opportunities and equal treatment, including equal pay, and my progression was entirely about how good I was at the job. I remained conspicuous however, and in some ways my performance was more visible than some of my male colleagues, who could hide any mistakes more readily.

I stayed with Marconi for 18 years, eventually leaving in early 1988 when the management regime changed and it became clear that the rapid progress up the career ladder that I had made until then was going to come to a halt, purely because the new Managing Director did not believe in having senior women in the Company. During that time I spent around three-quarters of my time on the New Street site, with brief excursions to Billericay, Waterhouse Lane, Writtle and the Research Station at Great Baddow, but to me, New Street *is* Marconi's.

Building 720 loomed large in my apprentice days. The canteen was at the top, there were offices on the first floor, an assembly area for large transmitters on the ground floor, and the apprentice training machine shop in the semi-basement – 'The Pit'. There we learnt the basics of turning, milling, planning, welding, sheet metal work etc, and the whole place had a unique perfume of cutting lubricant, which hung around one's clothes and hair. I never did grasp arc welding, but was up to pass standard in everything else, and much to the annoyance of the boys, came out top in turning. The method was that we were taught the basic techniques, then given exercises to do to show that we had grasped the techniques, and if you managed to finish all the exercises in the time allowed, were then given the opportunity to make small items which we could buy for the cost of the materials and take home. I took home a small brass and copper jug, aluminium perpetual calendar, and brass boxes that I had made, as well as making my own centre punch and scriber for later use.

The canteen was reputed to have the largest unsupported reinforced concrete roof in Europe and was a vast open area with the serving counter along one long side, where you queued to collect your meal. About a third was curtained off for the managerial grades to eat in – I think it was waitress service, but I never ate there. Certainly, it cost more to eat behind the curtain. There was also a Company Shop at one end for many years, where it was possible to buy GEC group goods at a discounted price. Later, the shop branched out into selling other goods that were of particularly good value – I am still using a set of Sea Island cotton sheets and pillowcases that are of fabulous quality. Schreiber furniture was part of the GEC Group at that time, and it was possible to buy their furniture at much discounted rates – a godsend when setting up house.

At the front of the site, facing New Street itself, is the original 1920s building, which housed the Publicity Department, the Surgery, Security, the Car Pool Administration and the Oak Room. This last was the Executive Dining Room, where senior management could eat every day (but I don't think did), or lower grades could bring external visitors and customers. I ended up eating there quite a lot when I was a Sales Controller. The food was definitely better, it was waitress service and there were cloths on the tables. You could even have a gin and tonic before lunch, while the rest of the site was very 'dry' – possession of alcohol on site was a sackable offence.

Running back up the site was a two storey 1930s building, the ground floor of which was an enormous production area and machine shop, and the first floor contained Purchasing, Contracts, some of the Sales functions and Personnel. The first floor corridor was offset so that one side was a row of cellular management offices, while the other was a series of large open-plan offices. When I first worked there, there was also a typing pool, where letters for the outside world could be taken to be typed up by rows of women sitting at very modern Olympus electric typewriters. The only problem was that they tended to disappear, never to be found again, or come out as gobbledegook, if your writing wasn't too good. A minimum of 24 hours turn-round time had to be allowed unless you were very senior.

Also on the first floor, but tucked away from the offices was 'The Winding Shop' – the Marconi equivalent of the Gulag Archipelago. Being assigned to the Winding Shop was one stop before disciplinary measures – I was never sent, I hasten to say. A group of about 50? women sat at machines that resembled a cross between a lathe and an old-fashioned treadle sewing machine, winding the precision coils that were used throughout Marconi equipment, not just Marconi Communications but for other GEC-Marconi companies as well. They were on piece-work pay, so woe betide the labourers who failed to supply the materials on time. Many of the women had been in the factory since the Second World War and they were tough and very scary en masse.

One Winding Shop incident that was kept very quiet occurred at Christmas time in the 1970s. One labourer was middle-aged and very unpopular – he was a 'dirty old man' who kept porn to read at his breaks, and made vaguely indecent remarks under his breath. The factory closed on Christmas Eve, and late that evening, some of us who were in supervisory roles were called in to undertake a detailed search, because his wife had reported that he had not come home, and checks of the hospitals etc had failed to find him. He was eventually found locked in a cupboard, stark naked. Apparently, just before knock-off time, a coat had been thrown over his head, he had been stripped and locked in. He couldn't identify who had done it (although we had a pretty good idea), and he never returned to the factory. The culprits were never found, but I guess the Company had to pay a fair amount of compensation. We hoped that someone would have eventually have rung Security and revealed his whereabouts……………….

The sheer size and complexity of the site meant that security was a nightmare. The 1970s was the time of the IRA bombscares, and we had to evacuate a number of times after coded warnings were received, although no bombs were ever found. Although the majority of the equipment produced was for communications, a proportion of it was for the Ministry of Defence, and hence we were seen as a defence contractor and a legitimate target. When alarms were sounded, everyone had to move outside the perimeter of the site, and then when the declared deadline passed, the supervisory staff (of whom I was one by then) went back in, to undertake a search. We became quite blasé about the whole process, although with our fingers crossed, because the number of boxes, containers etc meant that the opportunities for a small to medium sized device to be concealed was huge.

The risk was complicated by the fact that many of the labourers were of Irish extraction, and from both sides of the religious divide (and the Irish border). An example was in the paint shop, where equipment such as transmitters were sent for custom painting prior to assembly. There were two labourers, both of whom were about 60, one from the south (and Catholic), and the other from the north (and Protestant).

Both were functionally innumerate, although this did not stop them from having a very good understanding of odds at the betting shop. As Charge Hand, I was responsible for ensuring the smooth flow of work through parts of the shop, for which I needed the cooperation of both labourers, neither of whom would speak to the other, and I probably learnt more about the management of challenging people in that job than any other. During the period I was in the Paint Shop, the Company decided to move over from cash payment of staff to bank transfer, with the inducement of an extra week's pay for everyone who agreed to the changeover, and as the most junior Charge Hand, I was given the task of helping the manual staff though an understanding of how a bank account worked; most at that time did not have a bank account, and didn't particularly want one. Incidentally, I was the first woman Charge Hand in one of the 'male' areas of the factory since the Second World War, thirty years earlier.

The Paint Shop itself was a somewhat rickety building at the very back of the site, which had been thrown up at the beginning of the Second World War, and was something of an anachronism even then. It had single thickness brickwork, and was freezing in winter (due to the ever-open doors to let the work in and out), and hot the rest of the year, rising to unbearable in the summer of 1976, due to the tunnel ovens having to run to bake much of the paintwork. Although there was extraction, and the Company was very committed to making sure everyone wore the proper breathing apparatus and protective clothing, in retrospect, it certainly did not meet current-day standards of Health and Safety.

Many of the buildings at the back of the site were equally ramshackle, built in the early days of the war to build radar equipment. The Printed Circuit Process Plant, in which both I and my husband worked for quite a few years, spread over 5 or 6 of these buildings and a lot of time was spent moving between them if you were a Production Engineer, or maintaining the equipment, as I was. The equipment inside many of them was state of the art, but the buildings themselves were crowded, and for the most part single storey – a really poor use of space. Over the years, I was involved in three bids to demolish them and replace them with much more modern, well insulated and efficient accommodation, but we never managed to meet the investment criteria of Head Office at 1 Stanhope Gate, reigned over by Sir Arnold, later Lord, Weinstock.

Other characteristics of these buildings were the remnants of camouflage painting on the roofs and outside walls, which had been painted to look like a church and residential buildings during the war, and never painted over – just faded with time and weather. Three of the buildings also had pretty vast cellars, which had been used as air-raid shelters by the surrounding terraces as well as the factory, but were pretty much unused, although there was some storage of large items underneath. Between the Machine Shop and the back of the site was Marconi House itself – a four storey 1930s, semi Art Deco building with local Crittall windows. On the top floor were the management offices, including one which was called Marconi's office, and which contained some interesting artefacts, although whether he actually occupied it was seen as debatable. [He never did]. On the ground floor there was a very pleasant circular entrance hall, off which was the Telex Room; before fax and later email, the only way of sending quotations to customers and receiving instructions which were faster than post was telex. At one time when I was in the Printed Circuit Plant, I would be going to the Telex Room 6 or more times a day to deliver and retrieve messages to and from our main customers. It kept me fit, as the site was big and on a slope.

Against the back boundary of the site there was a long two-storey white building, built at the same time as Marconi House, which had Development Labs on the ground floor and a large drawing office on the first floor. I was told that in the 1930s there had been an open recreation ground between this building and Marconi House, but this had been built on during the War. Again, I spent much of my time as a Production Engineer walking to and from this building, consulting Design Draftsmen about solutions to production problems. No wonder I was slim and fit!

There were other eccentricities about the way the Company was run. One of them was the very, very small number of people who had Company cars. The Managing Director had both a car and a chauffeur, and I think the Production Director also had a car (but not a chauffeur), as he was responsible for a number of different sites. But everyone else had to have a car from the Car Pool, and making friends with the Managers of the Pool was an essential precaution if you needed cars often. When I was Sales Controller at the Printed Circuit Plant, I and my three Sales Engineers would have three or four cars most days of the week, but we still had to use the pool of very, very basic estate cars, and we were meant to return them to the pool every night – no keeping the car at home overnight unless you could show you were setting off at the crack of dawn the next morning. Initially, they didn't even have car radios to enable you to keep track of traffic hold-ups, but eventually, our pleas were listened to, and the next generation of pool cars did have radios. Our customers stretched from Kirkcaldy in Scotland to Plymouth and Cardiff, and we put on about 100,000 miles per year between us, but we still didn't get our own cars.

The New Street site was huge, and I believe that at its peak in the 1970s, perhaps 5five thousand people were employed there – In contrast, the evening 'twilight' shift, which varied between 5.00 pm and 10.00 pm, and employed mostly married women was about 1,000 strong, and the night shift, which was confined to the continuous process plants, such as the Printed Circuit Plant and Plating Shop, and the Machine Shops, employed a few hundred. As a Production Engineer responsible for the introduction of new machines and processes, I very occasionally had to work a night shift, and I hated it. I could have got out of it, because the Factory Act of 1878 (?) was still in force in regard to women not being allowed to work after 10.00 pm, but I never wanted to play the 'female' card, and instead was granted an exemption certificate, where I basically signed my rights away. However, liberation did not change the fact that the site at night was a very ill-lit, spooky place, and the men that worked permanent night shift a breed apart.

I went past the New Street site this week on the train to Colchester, and to see it deserted and overgrown is incredibly sad. It is hard to imagine it as it used to be, a location where I spent many happy years.'

Barbara M Stephens

Brian Wilson did his apprenticeship at Marconi's New Street. At one time he worked in the Instrument shop Section 16 (wiring and assembly) close to his Uncle George Wilson who at the time was working on a lathe in Section 15 of the Instrument Shop.

'I went to Barcelona in September 1960 - the dates were 24th September to 7th October representing Marconi's in the International Apprentice Competition. I am very proud of the fact that there were 5 apprentices from Marconi in a team of 30 that represented Great Britain. We all had to do a trade test during the first week, before the second week touring, ending in Madrid where the prizes were presented by General Franco. I was the senior centre lathe turner and had to make a crankshaft on a very old Russian lathe! We were kitted out with grey trousers and a black blazer which had a Union Jack badge sewn on the top pocket. Cannot remember where everyone came from while over there but unfortunately we didn't win.

Before the above event I had gained first place in the Senior Turning contest held by the City and Guilds of London Institute. Four other Apprentices had gained first place in their sections and were representing Marconi's, the others came first in Junior Turning, Senior Milling, Junior Milling and Senior Machine Designing Draughtsmanship.

The Marconi Magazine article at the time wrote: 'That five apprentices walked off with first place awards in the face of keen competition from apprentices of British engineering firms of world-wide repute is a cause of very real pride to us, especially as the panel of judges commented that while the standard of all the competitors was high, that of the winners was outstanding'.

At the time it was 'big news' locally as Marconi had sent eight apprentices to the International Apprentice Competition UK eliminating round and getting 5 first places was excellent.'

Brian Wilson

'I joined Marconi in 1962, did an introductory course, at Marconi College before b'seing sent to TV Test at New Street, working for Arthur Fisher. Then worked for Dave Parkinson at Rainsford Lane, which was an offshoot of Pottery Lane, before all TV went to Waterhouse Lane. There was an annual 'Open Day' and I volunteered to man a Mk III camera on the top of the OB van. We were not broadcasting, as such, but only for demonstrating purposes.

I picked up a rather interesting female, with rather nice figure and short skirt, and the Director, a friend of mine, said 'follow her' which I did, and as a result, went to an angle that was not designed for the camera shot, and I fell off the top of the van. All that was hurt was my dignity!

From WHL, I transferred to Building 46 in New Street, Transmitter Dev, working for Doug Bowers, Joe Sutton, and Ron Bradbrook. Started on a new design of HF Transmitter, the 300kW B6124 and built a prototype, which in the early stages of development, was sold to Nigeria in 1987. As a result two engineers from Nigerian Broadcasting were sent over for a training course in which, yours truly, was to give a lecture on the Transmitter Memory unit. After an hour, I concluded with 'any questions gents'. The immediate response was 'What time do Woolworths close?'

Later, I had to travel to Nigeria with Johnny Whatson, to finish the development, which proved impossible, due to frequent power cuts I had many happy times in Building 46, until I left for America in 1991, and I do have many amusing anecdotes including the time that a water load burst and the floor of Building 46 was awash with an inch deep of water. This resulted in 'comments', when visiting Marconi's.'

George Maclean

CHAPTER SIXTEEN

1970 - 2000

'**Late** 1976 and after finishing my first year as a Technician Apprentice I found myself doing my first 3-month secondment to a real department at New Street - the Automatic Test Department under the benevolent supervision of Dave Kingsberry. Who can forget sitting around the Myriad 2 computer with Alan, Dave, Tony and the gang at Christmas, eating Mince Pies, whilst listening to the machine play Christmas carols over its audio monitor!

One of my enduring impressions from my early days at New Street was the quality of the engineers. As a new graduate you arrive from university as a self proclaimed Master of the Universe – God's gift to the Company. Boy did that impression get dispelled quickly! After finishing my student apprenticeship I had planned to join Digital Equipment Development. A phone call at home from Brian Skingley, then Technical Manager for Space and Microwave Division, put me right and I found myself a member of the System Studies team housed in a small office in Marconi House. The team consisted of Rodney Braine, Mehdi Nouri, Tilak Artheniaka and me - did I feel a dunce in that company!

One of the least popular people at New Street had to be Andrew Glasgow OBE (Other Buggers Efforts), our one time MD. Boy did we laugh when one of the windows fell out of Building 720 and landed on his new Jag!

In the early 1980s Space and Microwave Systems Department was located on the 4th Floor of Marconi House and, looking North, overlooked the long overgrown site of an the old Brewery. One afternoon, presumably after a few drinks in the Old Ship pub, a young couple were spied making their way through the undergrowth. Completely oblivious to our excellent vantage point the couple got down to 'business'. Word soon got around - I have never seen so many people with their faces pressed up against an office window. After their performance we opened the windows and applauded!

In the 'old' days (1970s) I remember New Street as being full of characters - one was 'Eric'. 'Eric' had worked in Broadcast sales in the 1960s and covered East Africa. 'Eric' discovered that getting around such a large patch was time consuming and far from easy.

He, being a resourceful sort of person, came up with a solution - he bought a small airplane on expenses and learnt to fly. After his tour of duty he sold the airplane and credited the proceeds back to his expense account!'

Rick Potter

'I worked at MCSL from 1974 to 1990, though from 1974 to '85, this was at Specialised Components in Billericay, where we made Ferrite, Dielectric material and hi/low power microwave devices in waveguide and coax. During my 11 years there, I went from paint-Sprayer, to Contracts officer and then to Export Sales Engineer - though by the time we met, I was a Senior Buyer, for Dick Mays' Procurement Team, in Marconi House, buying material for Earth Stations.

During my time with Comms, I obviously worked with people who had been there for all of their working lives and who still told odd stories, of the 'old days'.

At Billericay, we had a chap in the workshop (retired but at 70, was still doing two days a week), who had worked at New St, before the war and recalled one MD, who was a former Naval officer (Admiral), who treated the place as a ship. Each piece of brass and all the fittings had to be polished on emery, prior to despatch.

I recall working with people from Baddow, who had known Chris Cockrell, who had to leave the Co. to invent the Hovercraft. I was always told horrible stories of the Winding Shop (cow shed) and advised that it was not the place for a young chap to be, on Xmas Eve........

Don't forget the tale of Gordon Hancock (was General Manager) who bought an aircraft on expenses, while Project Manager at an installation in Africa. '

Martin Gebel

'Ah.... the famous aeroplane story – I knew Gordon well and he sometimes (but not always) denied it. He also sometimes denied that it was Africa that the plane was possibly bought on a Company cheque, fuel might have been put on expenses and that the plane was eventually sold for a profit. So that clears that one up then.'

Tim Wander

[Gordon confirmed the story about three years ago – it was actually a Nigerian VHF Installation c1947-49. Outstanding!]

'**Building 46** was really bizarre.
Anyone recall John Temple, who did the inductor designs for the 50 kW Doherty, computer-calculated?
The Power Supply House was unbelievable.
Full of geriatric mercury-arc rectifiers. And the roof leaked!

One time I decided to play a trick on Ron. In the Crypt under 46, I discovered an ancient piece of test gear. Basically it was a 4 foot ebonite stick with a huge neon lamp on the end. I fixed it up with some extra nonsense bits and pieces and when Ron came round in the morning to see what I'd done to 'my' 50KW since the previous day, made sure he discovered me with the doors open, 'measuring' something with my device. I had him going for a few minutes until he blew my technical BS out of the water!

I decided to build the TX with open-mesh panel sides, so I could see everything from all angles. Ron's was very pukka with super-duper grey hammer and so on.

My idea worked really well in tracking down the cause of melting components etc. until one day the techie was sitting eating his sandwiches next to it, and I had made a mod that failed, resulting in a large cap exploding and spraying ceramic and gunge all over his lunch.'

Clive Warner

'I worked on a 300kW B6124 for the Swiss - I had to design an auto-tune function to accommodate the thermal expansion/contraction of the aerial.

I'd moved on before it was installed, but as far as I know it worked in the field as it did in the lab. It was triggered by an increase in VSWR and it made small retuning steps until the VSWR dropped back inside limits.'

Jez Cunningham

'**This** is what my failing memory tells me:

B6124 was 300/250 kW for Nigeria (Prototype) Swiss PTT (Swartzenburg), VOA (Woofferton and Greenville) and Rampisham with class A/B Modulation B6125 was 500kW version with Dual pulsam. Two Grid Boxes. Maybe this was Greenville N.C.

B6126 was 250kW for Skelton, Daventry and I believe BBC in Africa (Seychelles). I think this was single Pulse Step B6127 was 500kW I think that this was for Rampisham instead of 6124 B6128 was, all singing, all dancing 500kW with auto tune, Pulse Step, filtercon panels and stepping motors on the servos for VOA, Morocco, Thailand and Sri Lanka. Think that there was a medium wave version for Botswana. There was also a LW version for Droitwich.

George Maclean

A Welcome Visitor

'**This** took place at New Street in the late sixties or early seventies. I was not one of those actually involved in the drama, but was working in TV Test at the time. A young African TV engineer arrived at the front entrance one day, having come by taxi from Heathrow. There had been no previous correspondence with the Company concerning his visit and he spoke only French so it took a while to discover the purpose of his visit. It transpired that he had come from one of the small West African states (I can't remember exactly which one) in the President's private jet, bringing with him the entire complement of TV cameras (three in number) from the country's one and only Television Station.

The cameras were Marconi Mk.V Image Orthicons and none were in working order. He had orders not to return until they were all fully operational. I understand that the TV Service was left to run on a small continuity studio and a couple of Telecine machines. I must say that the Company pulled out all the stops with full co-operation between Broadcast Division, the Works and TV Test. Transport was despatched to Heathrow to extricate the equipment from Customs and, after an initial assessment by TV Test, wiremen and fitters worked overtime to strip down, repair and rebuild the cameras and their associated control units. Finally TV Test put them through a rigorous re-testing programme. As I remember it, the whole exercise was completed in less than a week. While all this was taking place, Broadcast Division arranged accommodation and hospitality for the young engineer. I know nothing of the financial side of the story, but I suspect he had a blank cheque from his Government to pay whatever it cost to get them working again. I hope this article may prompt others who were actually involved in the episode to contribute their recollections, and hopefully, correct any inaccuracies in my account.'

Charles Boyton

'I have worked for Marconi/BAE systems for 30 years on and off, but in the late sixties I was an apprentice at Hoffmann's, and I remember the fierce rivalry to obtain the best Carnival Float for the town Carnival.'

Terry

'I worked at the (originally Marconi's when I joined it) Research Centre in Great Baddow, but had reason to visit New Street occasionally. One of the times I was there, we were in some anonymous part of the site - I think it was just called 'building 46' or something like that. Anyway, just plonked next to the stairs there was a tall display cabinet (or fixture - might have been just the original equipment housing it had been in, but it had glass parts you could see in through anyway), containing what I understood to be (there was probably a label on it) the transmitting valve from the (first, I assume) television transmitter somewhere: Alexandra Palace I think. From what I remember, it was about the shape and size of a bell jar; anyway, everybody's idea of what a big valve would look like (definitely lots of glass in it, not all metal). It seemed to be totally ignored by those around it, as I suppose happens to something you see every day; I remember at the time thinking that I hope an object of such significance doesn't just get thrown out sometime.'

J. P. Gilliver.

'I originally started an apprenticeship at New Street in 1967 but left after a year for a better job and prospects with Essex County Council, Highways Department where I served the next five years as a Draughtsman / Technical Assistant.

In 1974 I returned to Marconi's and was employed briefly as a Sales Assistant. I then moved into Supplies Division, located on the first floor of Marconi House, becoming a Material Scheduler, assisting with the supply of components and hardware for the Telecine project. In 1976 I became a Draughtsman in the Standards team where I was mainly responsible for producing component outline drawings and material specifications to be incorporated into a new range of in-house catalogues. These catalogues were used by engineering to aid selection of components from a standardised / restricted range to enable costs to be reduced by the power of bulk purchasing.

Later in my career I was appointed the role of Standards Engineer, Mechanical Components and Hardware, still with an active role maintaining all the purchase specifications / drawings. However, in 1996 sometime after the Italian company, Marconi Spa had taken over, I was made redundant.'

Richard Puttock

'To pass the time at lunchtime we used to plot the position of the shadow of the sun on the roof tiles outside of the Tool Room Drawing Office / NC programming area on the top floor. We could see the true sinusoid on the graph and could predict equinox quite well. This was instigated by Doug Robinson of the Chief Production Engineers Dept.

David A. Johnson
Production Engineer, New St 1977 - 1980

'After 3 years in the RAF and 4 years at Mullard's, I joined Marconi's Wireless Telegraph Co in September 1963 working in Radar Test at New Street with George Jacoby and Herbert Ecclestone. I stayed at Marconi's until my retirement in February 1998, apart from 6 months at GEC Coventry in 1985. In the 34 and 1/2 years service I held about 12 different jobs. (I learned to keep on the move when in the RAF).

In the mid-1960s I worked at Baddow Research on 'Fur Hat', testing equipment for the early warning radar system for NATO, which was installed in Sweden. From there I moved to Beehive Lane then Widford, as Section Leader testing Myriad, Marconi's first venture into real time computers. Myriad One was as big as a desk, and had a whole 4,096 x 24bit word ferrite memory. Three Myriads were used at West Drayton for Air traffic Control for many years. Final test time was eight weeks, including heat testing in a polythene and wooden tent, which we called 'The Tardis'. This last test was called a 'schmoo test' and determined the working characteristics of the ferrite store. I was also responsible for the specification of the electrical installation at Widford site 2.

Later I became Test Section Leader in charge of System 4/30 computers, which were commercial computers made under licence for English Electric Leo Marconi. In 1969 I was promoted to Chief of Test Engineering (under Owen Ephraim) responsible for maintaining automatic test equipment used within Computer Equipment Production Division, which was managed by George Williams, and part of Works Manager Ray Stiles' R&D Workshop empire.

In 1972 it was decided to amalgamate the Automatic Test Equipment (ATE) activities of the Company, and I was put in charge of the combined testing, programming, maintenance and development functions with a team of about 30, based at New Street. We used Marconi Instrument's Autotest, plus 2 special ATEs called 'Comet' and 'Cobalt', which were developed by Brian Aldred & Geoff Collins, who were in Ron Kitchen's group, initially at Widford Hall, then at Baddow. Colleagues from that time included Dave Kingsbury, Peter More, Tony James, John Chiles. Also, in 1972 I had my first contact with Marconi Italiana, when I re-engineered the testing of Lincompex from Genoa methods using Siemens ATE to a New Street approach using MI test equipment.

In the later part of this phase Tony Wickens was in charge of the Test Equipment group, but in 1979 I applied for and got the job of Test Manager, HF products, under Cyril Marshall and Sid Poplett. This promotion was based on my computer knowledge, related to the successful launch of the MFT2 family of HF 1041, H1141 amplifiers, H1541 Drives and H2541 Receivers. In the early 80s, when Herbert Ecclestone retired, I combined Quality Controller with my Test Manager role.
Would you believe it, in 1981 there was another reorganisation, which combined HF into Radio and Lines Division. I became Test Engineering Manager, with a million pound budget for ATE for PCM and Kilostream. MI System 80 was used, plus functional ATE from MI, and duplicate programmes, fixtures and ATE were produced for Altec in South Africa.

I managed the design and production of ATE for Automatic Cross

Connection Equipment which was used to switch voice streams within a 2 Mbit/s network (ACE). This project was a big challenge and cost half a million pounds. It involved managing 'softies' (software engineers), and we all know what a problem they can be! I had the help of Richard Tagg, Trevor Mann, Norman Ibbotson plus Jack Dyer. My team pioneered the use of the BBC computer in Marconi Communications and these simple but powerful computers were made into low cost ATE controllers.

In 1985 I was 'head hunted' and transferred to GEC Telecom working for Brian Aldred as Test Engineering Manager in charge of 50 people, but the difference was that design for testability was mandatory in Coventry, optional in Chelmsford. I enjoyed this job for 6 months but my family couldn't make the move, so I came back to New Street. Jim Cole kept me busy as part of the Computer Integrated Management team, devising the strategy for Computer Integrated Testing. I worked with Paul Jackson and Gary Jenkins, and during this phase I wrote my first book, entitled 'Computer Integrated Testing' published in 1988, based on the pioneering work I was doing at the time. In 1990 I accepted an invitation to lecture on Computer Integrated Testing in Singapore and I was impressed how keen their electronic engineers were to use all the latest technology (unlike Chelmsford where it was hard work changing attitudes). A later book was published in 1994 called 'In Circuit Testing'.

In 1987 I accepted the challenge to become a Manufacturing Project Manager, on the RC630 project for British Gas, then on RC690 Scimitar radios for Turkey. I worked with Archie Goddard, Lee Snook, Paul Jackson, and Keith Lamb, under Mike Steel.

1991 saw another move, as ILS (Integrated Logistic Support) Project Manager under Joe Bailey, working with Chris Elliott on the JORN project. ILS is concerned with manufacture-ability, testability, reliability, maintainability, so it aims to influence design for the benefit of the customer to optimise the use of the product through its whole life. My role changed to ILS Technical Manager when Bob Sykes came on the scene from Australia, and we worked happily together until near the end of that project, some of the work occurring down under with ex members of the Royal Australian Airforce.

In the mid-90s ILS was gaining recognition on many MOD projects so I moved off JORN and for my last 2 years before retirement I worked on secondment from New Street to Crawley on preliminary studies for a communication project for a new generation of Frigates. The MCSL team worked with engineers from Redifon, Thomson-CSF of Paris and Elmer of Italy. I was surprised how reluctant Marconi Management were to obtain as much business as possible from this project compared to the other companies, who were keen as mustard to grab what they could. I was told to look at the 'bigger picture'. '

Alan Buckroyd

'**While** all the hardware was built at New Street most of the software development for ACE was at Writtle. One memory I have is of the ACE advertising video that was made. William Woollard (Of Tomorrow's world fame) was to be the commentator. I had to give him some instructions. I had written a show crib and was talking it through with him and called it 'The idiots guide...' He took real exception to that and said '... I have two degrees...'.

During this video we had several of the hardware engineers pulling out cables and re-attaching them to simulate the faults - really funny to observe as they were squatting down behind the huge racks. But how else can you generate faults without a complete network.

One day whilst sitting in my usual spot beside the racks a sudden shout went out from one of the hardware engineers (Pete Botham I think) 'Catriona , That must be a software bug!' And lo and behold there was!! A huge black beetle running out from underneath the rack of equipment I was working on.

The first New Street development I remember was on PMR.' [Private Mobile Radio]

Catriona Potter

'The RC690 'Repeater Mobile' is a boot-mount transceiver with a separate control head unit. It was made in the 1980's by Marconi, Chelmsford and sold to many police forces and ambulance fleets. 'Repeater Mobile' comes from its ability to act as a repeater for the Police handhelds when the officers leave the vehicle. It is also classed a 'Wideband' mobile due to the wide tuning band without factory realignment. RC690 is capable of full-duplex operation with two antennas but may also be fitted simplex with one. I'm told about 3000 were were made. There exists several variants: VHF mid-band duplex 143-156MHz and a VHF low-band simplex-only 70-84MHz.

The RC640 'Repeater Mobile' is a boot-mount transceiver with a separate control head unit. It was made in the 1980's by Marconi, Chelmsford and sold to the Metropolitain Police (London police). It appears they were also used by ambulances in the West Midlands and railways in Iraq. I'm told only about 100 were made.'

Dave Mckay

Marconi New Street Manufacturing, c1986

**Visit To Marconi Communication Systems Ltd
Of HRH The Duke Of Kent**

'A fine morning in April (1980s) greeted me when I was telephoned at my home in the early hours 2-30am, with a request to come to the New Street factory to meet members of Essex Police.

On arrival I was advised that it was necessary for Essex Police accompanied by Special Branch Police Officers to thoroughly check out the factory and the route intended for the Duke of Kent to take when he arrives later in the morning. I duly accompanied the Officers who had sniffer dogs with them and we covered the entire route and many other places including the Directors Luncheon Club finishing around 6-30am. The Police Officers stayed on site whilst I retired to my bed for at least another hours sleep.

At around 9-00am I advised my Managing Director of the events which had taken place earlier and then prepared for the visit of HRH . A helicopter had landed at Waterhouse Lane on the English Electric Valve football pitch and cars conveyed the Duke of Kent and his attendants to New Street. Following introductions to the Managing Director and other senior staff including our Life President of Marconi Company Sir Robert Telford, I led the party through the factory, ending up at the top end of building 720 where the Duke would see some of the High Power TV and Communication Transmitters before ending up in the Directors Luncheon Club.

After some five minutes in building 720 one of the plain clothes Police Officers came up to me and asked if I could delay the party from going outside for at least five minutes. There was no explanation why. I advised the Managing Director to slow the visit down but he was not happy as the whole visit was timed almost to the very minute. I insisted and in the end he complied.

After the party had left the building on the way to lunch, I caught up with the Police Officer and asked him why it was necessary to delay the party in Building 720. He related the story. The night before the visit a hunger striker had died in a Belfast prison and the whole country was therefore put on an alert as trouble was anticipated from Irish dissidents. Whilst keeping a watch outside building 720 the special branch officers had spotted two men walking along the railway line which was behind building 720 and divided by a very high fence.

The special branch men decided they should take some action and rather than challenge these men from where they stood they decided young Police Constables should get over the wire fence and question these men. Apparently they went to the men's toilets and got hold of some duck boards, put these against the fence and clambered over the top. When they got to the men they challenged them and asked what they were doing. The response in Irish accent was 'Sorry mate but we were heading for this clump of bushes to have a pee'

The problem was over and two young Constables had partly torn uniforms from the barbed wire and HRH The Duke of Kent knew nothing of the event.'

Peter A.T. Turrall MBE

New Street Memories 1983 - 1997

'I joined Marconi's straight from University with a shiny new degree in Computer Science from Aston University in Birmingham. At University I had tortured and then even helped scrap the ICL 1904S computer using GEORGE III as an operating system. I fed it its last program loaded on punched cards, now museum grade technology. I was offered three jobs from University, with Racal, Phillips and Marconi's. Racal offered more money (£200 p/a!), Phillips several months in Holland training which sounded fun and a bit more money but then you ended up in Reading which on my only visit for an interview seemed like a modern wasteland. Marconi's offered 80 plus years of history and a chance to join one of the greatest names in the history of radio. I already had an interest in amateur radio and restoring old valve radios. It was really no contest.

So I reported in my shiny new shoes and slightly worn interview suit to the Computer Services Unit (CSU) of Marconi Communications Systems Ltd (MCSL) based at the historic Writtle site where British Broadcasting was born. Little did I know that over 30 years later Marconi and the Writtle story would still be with me practically every day, even though I left the Company over 16 years ago.

On arrival at Writtle I was shepherded into a small room. I was now a 'baby softie', officially a junior software engineer and the big boss was software guru Mike Tate. We sat in the small room and were given daily tasks coding software or testing small programs until gradually each of the 7 or 8 engineers was 'hired' into a project team. The biggest projects at the time were the ACE project (Automatic Cross Connect Equipment) a large multiplexor and its control system primarily for BT. MOFA was a message switch system for the Saudi Arabian Ministry and the ARC project, or ASWE Remote Control that had grown out of the ASWE Map Project and later became the Royal Navy's UKMACCS system.

Essentially the ARC project was designed to provide the Royal Navy with gapless HF radio cover within a 200 mile radius of the UK shoreline along with HF, RTTY, VHF and voice links. Even though the software was written at Writtle, every piece of the ARC project, the transmitters, power banks, 10 kW amplifiers, aerial switches, communications kit and all the control cabinets and hardware racks that housed our precious software were designed, built and assembled at the New Street Works, mostly in building 720.

The drawing offices, stores, project management, installation and all the other services to build the project also came out of just one site. In the main New Street building the works who were also busy building the many ACE cross connect units for BT and others, well over a 120 when I last counted and I guess somewhere in there the message switch system for the Saudi Arabian Ministry (known as MOFA) racks (and other AFTN Aeronautical Fixed Telecomms Network switches) were built as well, together with small things like the IBA TV transmitters for Channel 5, (having already shipped Channel 3 and 4 equipment) early satellite communications systems for the military, and Swordfish military radios and.... These were big projects and absorbed much of the Company and the New Street Works from 1979 until 1995.

So I was herded into the ARC project working for Dave Webber. It was a huge project, 30+ software engineers, 15+ hardware engineers and a cast of many hundreds back at New Street. The 19 inch ARC control racks had special in house designed processor boards based around the still fairly new INTEL 8080 processor (later Z80) and all the equipment was controlled via another dedicated control unit the

RPI or Remote Peripheral Interface Unit – it was this box that linked the control system to the transmitters, receiver, power banks et al. There was also a custom communication board (the QCI - Quadruple Communications Interface) that allowed the various remote sites (30+) to communicate over dedicated telephone lines as there no internet in those days.

The United Kingdom Maritime Automatic Coastal Communication System (UKMACCS) was a dedicated control system developed for the Royal Navy, providing a full remote-control capability over unattended HF coastal communication stations. Started in 1982, the control and management system was installed in 1985/1986 and became fully operational in mid-1987 and provided an uninterrupted service since that date until it was decommissioned in 2002. New Street built it all.

There were two separate control centres, a primary centre located in London and a secondary centre at a naval establishment in Yorkshire. The sites were duplicates and the emergency changeover software involving hundreds of pieces of equipment under control was a thing of beauty…..when we eventually got it to work. The HF coastal communications comprised many separate receiver and transmitter sites located in Scotland, Cornwall and other parts of the UK, some located in the remotest parts imaginable. I know because I went to them all, many times. The controlled equipment at each site included a number of HF drive units, high-power amplifiers, RF combining equipment, air cooling systems and an antenna exchange. A large number of HF receivers and an antenna exchange were located at each receiver site.

The control system provided full control of all equipment, at each remote site, from either of the two control centres. Indications from the station service equipment at each remote site, including fire and intruder alarms, were automatically monitored and a notification or any alarm condition was provided at the control centres. A purpose-built supervisory console was installed at each control centre allowing full control of all equipment by means of command inputs to menu-driven VDUs. The console also provided a control facility for an audio matrix unit, allowing the allocation of radio channels to various operators or services. Provision was also made for the monitoring and display of received signal strengths allowing the optimisation of reception on each channel by the selection of appropriate directional aerials. This was a great bit of software called the AD module – but I am biased because I designed, wrote and tested it.

On the 11th November 1987 the Writtle Site closed down and the whole of the MCSL site and personnel were transplanted to New Street. We arrived like wandering nomads on the top floor of Building 720 with its unique preformed concrete roof and took over what had been the top floor canteen. I had first gone there in 1983 to play badminton and we had to move all the tables and then wash the floors with sugar soap every time we played to remove the slippery canteen 'grease'. The new prefab offices and the new software engineering area was purpose built with suspended floors and ceilings on the top floor of 720. It gave lots of space but air conditioning and even heating would have been a luxury. The builders also carefully installed lots of overhead fluorescent tubes which were less than perfect for 50 plus Visual Display Units that made the building very warm during the summer months.

When we arrived at New Street parking became a huge problem and for months we had to park at the end of New Street beyond where the University is now. Attempting to park inside the New Street yard became a daily challenge with the security guards at the front gate.

After ARC we also provided a similar system for the Oman Navy and then the same CPU/RPI/QCI equipment was used to control the new generation of Channel 5 TV transmitters being built below us in 720. After time on the ACE and then the Kilostream IV Multiplexor Projects including development of the new DEC Vax based management system that could control both ACE and Kilostream equipment called the RENACE.

I then started working on a set of new and advanced projects for MCSL including a new Remote Distress Signal Unit and Monitoring System for the Fire Brigade. We had to fight daily for funding and development resources as successive MD's kept saying it wasn't a core business. This project was eventually transferred to EEV and is now in operational usage having saved numerous fire fighters lives. During this period New Street started working with Marconi Italy in part due to the takeover of GMCL by the Italian sister company. We then had a couple of Italian MD's who never really understood the English way.

Hidden away above manufacturing my team at New Street quietly pioneered the technology behind tagging vehicles and we worked long and hard to bring the next generation of microwave tag technology to New Street. At one time we led the world in developing and testing the microwave interface communications protocols and the team helped write several of the European standards.

It is a little known fact that if the main board of directors of GEC had not stopped it at the 11th hour (because it didn't have enough Telecommunications *content* and George Simpson didn't like the idea of Marconi helping the Government to *tax* drivers) Marconi and New Street would probably have won the London congestion pricing system, possibly the Dartford tunnel tag system refit and all the others that have and will follow. We spent over 18 months testing and trialling including installing equipment in the MD's Jag and driving it up and down the New Street yard while overhead equipment talked to it every step of the way. Competing companies via the EU commission came to watch as they didn't believe we could make it work. We even convinced the Government after six very long, cold and wet months at the TRL test track that the technology worked. Then the French took it all and I left the Company.

During my 17 years with the Company I lived through an amazing revolution in the computer industry. Every day brought faster processors and bigger memories, but the problem was that large projects had to be designed around the technology available when they were started. The ARC project took over seven years to complete, install and commission and the world had started to move much faster than that. That is why the last space shuttle flight this year still had a main computer with less processing power than the laptop I am typing this on.

In the early 1990s I did a stint in Manufacturing, working for Mike Steel, but the writing was on the wall. It was sad to see the last projects go through the once proud manufacturing lines, from memory LNBs for Polish Satellite TV and the display panels for Gilbarco fuel pumps. After that the long line that Marconi, Kemp, Ditcham, Round and Franklin had once walked down stood empty and abandoned. The once proud Company simply couldn't compete with Hong Kong, Malaysia and China and there had been no investment for ten years or more. Over 4,000 people lost their jobs, sometimes three generations of the same family worked there.

There was the Director's Dining Room at the front of the building, oak panelled and you had to be very important to eat there. I first went there as a new graduate, to meet the Directors. A great spread of food was laid out on the huge immaculate polished table. I was asked to pass the silver tray of sausage rolls, but I only picked up one handle on the strange silver serving bowl and promptly dumped all the rolls on the table. I was helped by Project Manager extraordinaire Gordon Hancock to retrieve the faux pas, he assured me that 30 years earlier it was he who passed the soup tureen and you could still see the stain the hot brown Windsor soup had left. Gordon worked with me on and off for the next ten years and taught me much about the Company and Project Management.

When Marconi Radar/Eastwood House was built it was fascinating to see them find the original Blacksmith Forges, Air Raid Shelter, and the huge concrete foundation block for the massive 450ft mast. Then a JCB even found the old Brewery vaults by falling into a huge underground cavern. The building of Eastwood House meant that the original buildings including the one that Dame Nellie Melba sang from were all swept unceremoniously away.

Sending apprentices to the stores for sky hooks, or a long weight was popular. My version was to send a junior engineer with an old 1970s NORBIT (a fixed logic circuit constructed of discrete components and encapsulated in epoxy) down to get it programmed by the Works. They always sent them back with a 'completed form'. I wish I had kept them but suspect they couldn't be printed here.

The ARC project was written in CORAL 66, a now long forgotten programming language soon replaced by PASCAL and ADA and now, no doubt, many others. The CORAL programs were written longhand on coding pads by the engineers who had already produced design specifications and the forms were then typed by the data prep girls. Often each program was typed twice by a different person and the files compared by computer - very rarely did two typists make the same mistake so the flagged differences or errors were easy to find.

Once typed and syntax checked the programs were compiled on the New Street VAXes, submitted to batch queues to run overnight. Each project had its own batch queue and each ran with the same priority. The VAX could run perhaps 7 or 8 jobs simultaneously and it was up to the System Mangers to balance the priority against the speed of response for the users. Several of the 'Data Prep' girls could type so fast that they would fill up the type ahead buffer on the Computer and have to wait for the machine to catch up.

A complete system rebuild or recompile for the ARC project would require 10 to 12 hours of CPU time on the main VAX cluster alone with no users to slow it down, so the last job every night was to load the batch queues up and let them run to produce assembly code and download machine code files for the morning. If only it was that easy. Typically each compile run would throw up several hundred compile errors and each one had to be fixed and that part recompiled in the morning. As part of this one program, the common area held every data point and variable for every program, essentially the systems database all be it a very early one. That was my world for two years but it meant I got to work with every part of the system. For a while I was privileged to have at home a rare 'dial-in' modem so I could talk to the VAX from home and continually monitor and resubmit jobs from home making maximum use of computer time, especially over weekends. I burnt enough mill time to make NASA proud. Today its standard practice, in 1984 to dial into a massive mainframe computer from home was a rare and special privilege. It also probaly helped that I married one of the system managers…

Eventually correctly compiled code would give a machine code output file that could be programmed in the New Street Works (or by the engineers downloading) into EPROMS (Electrically Programmable Read Only Memory) that were erasable under UV light (32K, 64K that is only 8K x 8 **bits!** and later the 27128 and even 27256 chips but the latter proved expensive and unreliable). The Company did invest in some MICE units, which were microprocessor emulators that plugged into the Z80 socket and we could actually control the operation of the software and even step through each instruction one by one. You could set break point and builds traps and actually watch the soul of the machine operating. How many of today's game playing PC experts get to play with the actual bits and bytes and actually watch a processor executing binary instructions.

We became experts in reading machine code, the hexadecimal representation of the actual binary code and could write pages of patches in machine code and edit the actual chips (by loading into a programmer editing on a key pad and then re-blowing a new or erased chip) rather than recompile and reload which took hours if not days. I suspect these skills are now all lost.

It was a salutary lesson for any new software engineer to sit in building 720 in front of an MFT H1141 10 kW HF amplifier or a massive air cooled B7546 UHF pulsed Klystron TV transmitter for Channel 5. By changing just one binary bit, say from FF to FE, i.e. 1111 1111 to 1111 1110 in one memory address would cause a massive TV transmitter to switch on. I remember one engineer who got it wrong attempting to instruct a 10 kW amplifier to turn on and off, continuously, probably 3,000 times a second. The huge amplifier, as ever stood up to this abuse, but it did seem to whimper slightly.

New Street made the best transmitters in the world, both TV and radio. The MFT (Marconi Fast Tune) H1141 10 kW amplifier was an especially robust piece of kit. They have been dropped off trucks, transporters and even an aircraft, hit by lightning (many times), used as communal toilets and been attacked by everything including armed troops and they just kept on working. I have even seen them work after someone forgot to plug the aerial in. In fact they were so good that they were copied and even cloned by Eastern European Companies.

The Marconi Fast Tune (MFT) range of equipment was an important product for the Company but in addition to the drive and 10 kW amplifiers there was also a 50 kW L.S.B amplifier, a 1 kW amplifier and a transceiver. But by far the most important part of MFT2 was the remote Computer control system that goes along with the basic hardware; this system allows for automation in service selection which includes frequency, mode, audio source and antenna selection, along with any other aspect that needs controlling. It gave me a job!

In September 1978 Marconi's installed B6034 50 kW MF transmitters into Brookmans Park providing the BBC with Radio 1, 2 and 5 live on 909 kHz. These were only switched off at 1.0 a.m on 29th January 2009.

Throughout the 80s and 90s the software engineering departments was populated by two types of engineers. There were the 'permies' and the 'contractors'. These 'guns for hire' formed at times 80% of the engineering teams, could earn up to ten times the money of the permies. Yes they could be hired and fired at a moment's notice, but if they were any good and many of them were very good, they stayed and stayed. Within 2 years, out of all the baby softies I joined with I found that I was the only one left who hadn't gone contracting.

One amusing discussion and exchange of letters with the Royal Navy was a piece of equipment that failed in the wilds of Cornwall. It was found that a mouse, or a family of mice had made the main circuit board their home and, of course, every home has a toilet. The board had struggled on until it could take no more and it was returned or rather salvaged for repair, not a nice job for an apprentice who had to sit outside in the yard to re-solder it because of the smell. Not many spares of this one were available. Of course, the arguments over who paid for it all centred around the origin of the mouse. Was it in fact a conscripted military mouse who took up residency after the system was operational, a civilian mouse that had entered the base before official hand over or indeed a sub-contractor mouse from Chelmsford that had been delivered with the kit!

In the same vein at a receiver site in Cornwall the local rabbits took a great liking to the coaxial feed cables. We spent months pushing new types of cable down rabbit holes and then hauling them out each morning to see which types they liked the least. Luckily, it was agreed that these were indeed military bunnies.

Every year Chelmsford would host the Real Ale Beer Festival that kicked off on a Friday, the strict instruction, especially to all the younger engineers, who bought in their own tankards was to go at lunchtime and don't come back on site. Of course, my phone rang like one Friday evening. It was the Security Manager, Jim Jones, telling me that one of my team was in the New Street Yard, serenading (with a guitar) the MD's Jaguar parked as always in the first spot outside Marconi House. Of course he had no shirt and was wearing union jack shorts – perfect. We rapidly ushered him into a taxi.

But through it all, the deadlines, the crisis, the shouting and the panic there was always an amazing team spirit. People were proud to work for the Company. They gave it their all and often worked ridiculous hours at the expense of friends, family, relationships and even health to get some now long forgotten project out the door or into its next phase. The sad fact is, despite this dedication, New Street and in fact the whole Company continually missed or squandered the chance to pioneer new technologies.

At its very core, until the day someone turned the last lights out in the New Street Works, Marconi's was still a traditional British Company, with the very best and very worst that phrase has within it. New Street had some of the best engineers, designers, draughtsmen (whose ranks were unfortunately decimated within months when CAD/CAM came and it took several trucks to take the beautiful 1930s counterweighted drawing boards to the scrap yard), wiremen and fabricators. I worked with some of the best installation engineers who could seemingly overcome any problem in the remotest part of the world.

But by and large the Company Directors were remote and often aloof, locked on the 4th floor of Marconi House and protected by outer offices full of formidable secretaries. In the days before email, boy did we write a lot of memos and reports and CDR's (basically begging letters for funding) but meetings were a rarity and often then only when something or someone had really screwed up. Usually you heard nothing, but towards the end of my time there the 4th floor was overrun by accountants, but even they couldn't add up the losses fast enough. The smoking ban meant two or three 'Bus stop' shelters were erected in the main Marconi yard, one alongside building 720. Addicts puffed away out there in all weathers, but luckily two of the senior Directors were heavy smokers and as my office over looked the 'Bus Stop' access and signatures became easier.

So over cautious Directors kept the Company out of the Computer, Integrated Circuit and Microprocessor revolutions. Marconi's unceremoniously dumped the mobile telephone business even though the Company had the lead at one point based at Waterhouse Lane. It sold Marconiphone and I think some 80,000 subscribers to a very small company just starting up called Vodaphone somewhere in Reading. I seem to remember someone panicked over the first analogue phone cloning problems. Who could possibly imagine that everyone would want or be able to afford their own *portable* telephone one day?

Apparently that well known and historical centre of microwave radio and communication excellence, that was Finland. I remember my first Company 'mobile' was an analogue transportable, we called it a 'luggable'. There were just three phones on site that had to be signed out and each call and its duration had to be logged in the phones personal call book.

Once upon a time Marconi's led the world in Computer Technology with the Myriad System, but every opportunity and request to enter the world of integrated circuits, memories and computers were rejected by the management. Of course we were told that the personal computer had no future because we had invested in VAX computers that could run 60 people. I once heard that the Board actually rejected an order for PC's because they were toys and Marconi wasn't a toy Company.... but there again MICROSOFT and APPLE seem to do quite well in the *toy* business.

It is true that the Marconi transmitters, amplifiers and power banks were amongst the best in the world, but they were nowhere sexy enough for the new breed of Corporate senior 'managers' who took over towards the end. They were considered Victorian engineering (even though that had built the Company) but starved of investment and development systems and sales petered out. The rush to sell the whole parcel to British Aerospace was almost rude. The next headlong rush into the embryonic and risky world of internet start-ups and telecommunications was even ruder. It seemed to me that the new breed of managers, who preferred jeans and leather jackets to business suits treated the once proud company as an experiment in corporate finance and when it failed spectacularly they walked away with golden handshakes worth millions. Not so the ordinary work force who not only lost their jobs but also in many cases retirement nest eggs built up with share purchases encouraged by the Company.

Perhaps I sound bitter. Far from it. My years with the Company were amazing, I travelled the world, met and worked with great people and was always fascinated with radio communication even though I was at first just a baby softie....and as I may have mentioned the Marconi story is still a part of my daily life.'

Tim Wander

'Tim - I will always remember the day in the TX Hall at Klacmc'kenny , if that's how you spell it [close enough!], when you [Tim] stated you had found the problem with the intermittent System failures down to a duff EPROM and you flung it down the TX Hall just missing me as I worked on the combiner. Then some time later found that you had no more EPROMS and had to find the EPROM 'rocket' you launched, straighten the legs and reprogram it to fit elsewhere.

I did find the problem some time later on, it was on that EPROM location the actual socket had 3 very short legs on it and were just about touching the circuit board most of the time, solder blob's comes to mind to fix it.'

Steve Wakerley
EX Marconi Installation Dept New Street

UKMACCS ARC Team
L-R: Ken, Phil Watkins, Tim Wander, Trevor Groves, John Mayne, Peter Hopp, Nick Keighley, Terry Worton, Graham, Dave Webber.

UKMACCS ICS3 1kW Power Bank

UKMACCS Systems Rack, RPI, QCI and lots of fans. In Oman I had to go to the local street markets and buy all the desk fans we could find to keep the equipment cool while waiting for extra rack mounted trays to be delivered. The system never cooked but the room got really warm...!

Marconi 1141 MFT 10kw Amplifier

'I worked there [New Street] in 93/4 as a contract programmer. They had been taken over by an Italian branch of Marconi. The first thing that happened was the supervisor took me aside and asked what it was like to be a contractor... The new owners had sent all the chairs away to be re-covered and then refused to pay the bill and so everyone was sitting on whatever they could find. There were no partitions any more because the new boss was too short to see over them and had them removed. The code I was working on had been written by students on work experience who had now gone back to university, leaving a mess that wouldn't compile small enough to fit in the ROM of the digital Navy radio system they were developing. Fond memories! The canteen was excellent - chips with everything for the workers and pudding included

- for a quid. You could go and see some of the original transmitters in a museum area. I believe the men's toilets were original too... '

Nick Pettefar

Judith and the VAX Computers

'**I** joined Marconi Communications at Writtle as a Data Prep Supervisor in 1982. I learned many new skills as a Computer Operator.... this was very new to me because I had been a shorthand/audio typist in London for 3 years! I had taken all my exams on a manual typewriter – and my bosses (Graham Killens and Mike Tate) told me how easy it would be to type on a console connected to a computer. And how right they were!! I saw the paper tape machines go and the Myriad Computer taken away... and the new VAX 11/780 computer arrive! I became a System Manager and was responsible for the preparing and keying in of engineering data into the computer and downloading data into EPROM programmes for the EPROM production in the Manufacturing Department. It was during this time that I met my husband, who arrived at Writtle as a software engineer and used the VAX computer!

After five years I was given day-release to study part-time at the Polytechnic of East London and c,ompleted my Diploma in Technological Science.

When the Writtle site closed down the Bureau was transferred to New Street where I continued to work as a Systems Manager. The computer room expanded and I worked in a team of three with Derek Morris and Paul Dickens. We maintained a VAX Cluster consisting of six 8600 series VAX computers, also system managing other stand-alone VAX machines during this time which supported other departments within the Company. The computer room was large and fully air-conditioned to maintain these large machines along with their disk drives and printers and they provided all the computer facilities for Marconi Communications. These machines cost millions of pounds, and we were fortunate to have one of the first 8600 machines when it arrived in the country.

This was state-of-the-art hardware/software and we had maintenance engineers that we called immediately any problem occurred. We had to regularly back up their data monthly over night as it was essential that the machines were kept running continuously.

Backups were done onto an evil TK50 tape drive that used a CompacTape Cartridge that contained 600 feet (about 183 m) of half-inch magnetic tape on a single spool. A cartridge could hold up to 100 MB. Yes that is MB. My keyring has an 8GB memory built in!

Any unplanned downtime meant loss of work and money for the

Company – and an awful lot of angry people! In those days the Company had around 6,000 personnel, and all computer work was carried out through these machines. We had names for some machines, to make life easier, some were named after the Muppets!

In 1992, everything changed. The New Street computers were moved to Baddow as I left the Company to have my first child. It was the end of an era.....

Personal computers on networks arrived, and the age of the large mainframe computer went. I saw them arrive, and saw them go, it only lasted for a decade! Computers have changed rapidly in this time of technological change... but I was pleased to have been a part of it.

Judith Wander (nee Paskins)

The VAX 11/780 is seen as the high point in DEC's engineering and commercial history, although its raw specifications, a 5MHz clock, 8MB maximum RAM aren't that different to those of PCs that came just 10 years later and by today's standards laughably small. However, the VAX machines achieved an actual performance level that belies those figures, supporting 30 or 40 users with ease. VAX stands for 'Virtual Address eXtension', as the computer used virtual memory and a very efficient I/O system to behave like a much larger system. The actual performance was around one million instructions a second, making one VAX MIPS an industry standard benchmark metric.

> **'The** VAX system at New Street would support over sixty users and run at least 8 bath processing queues. There was no sign of a windows environment – actually all the terminals the engineers used were 14inch VT100 type black and white (green and white!) but still the New Street Works designed and built world beating equipment and hugely complex software systems on a system with less memory then the laptop I am currently typing on.
>
> VMS was a fairly solid operating system. Unlike any version of MS Windows, it was possible to make VMS secure, and unlike most versions

of UNIX, VMS came secure 'out of the box' (provided you changed the default DEC field service and system manager accounts!). Every user and every file had its own limited access and control privileges, known as R,W,E,D (read, write, execute, delete) except for the system manager (me!) – I liked the fact that in a Company full of some of the best engineers in the business only I could type > set proc/priv=all that allowed me to do anything and go anywhere. Every user also had their own access/privacy settings for their accounts i.e. restricting who could view their information and files; the categories were S,W,G,O – System, World, Group or Owner – sort of a mini Facebook twenty years earlier.

Other things that modern computer users will find strange (like my children) is that local printers were very rare – all printouts were run on large noisy high speed printer machines about the size of a large freezer in the printer room – 'list-outs' on A3 size continuous lined green 'music paper' were folded and put in each projects pigeon hole – only a few people were allowed access to the computer room.'

Judith Wander

In the United Kingdom, the Packet Switch Stream (PSS) was an X.25-based packet-switched network, provided by the British Post Office Telecommunications and then British Telecom starting in 1980. After a period of pre-operational testing with customers (mainly UK Universities and computer manufacturers at this early phase) the service was launched as a commercial service on 20th August 1981. The UK Private Circuit Network provided an extensive national private circuit network transmitting at speeds up to 64KBit/s under Kilostream and 2Mbit/s under Megastream. The entire network used Marconi ACE sites (Automatic Crossconnection Equipment) and was controlled from Marconi VAX based network controllers in Manchester and London. All the equipment and software was designed, built and tested at New Street.

PSS was one of the first telecommunications networks in the UK to be fully liberalised in that customers could connect their own equipment to the network. This was before privatization and the creation of British Telecommunications plc

(BT) in 1984. PSS could be used to connect to a variety of online databases and mainframe systems. Of particular note was the use of PSS for the first networked Clearing House Automated Payment System (CHAPS). This was a network system used to transfer all payments over £10,000 UKP (in early 1980s monetary value) between the major UK banks and other major financial institutions based in the UK. It replaced a paper based system that operated in the City of London using electrical vehicles similar to milk floats.

Marconi's also developed the equipment and control software for Kilostream. Digital private circuits are dedicated, permanently connected digital voice or data circuits between designated sites. Available in a range of transmission speeds between 2.4Kbit/s and 64Kbit/s, they are ideal for linking low-speed local area networks (LANs) and supporting applications such as stock control, online access to a host computer, ordering systems and remote printing facilities for PC users at branch offices.

Kilostream became British Telecom's lower-speed digital leased-circuit service. Kilostream circuits are still available as a fully digital service, operating at 2.4, 4.8, 9.6, 19.2, 48, and 64 Kbps. There is also a service called Kilostream N, which offers speeds in preset multiples of 64 Kbps up to 1.024 Mbps. Kilostream is also available as an international service. Before broadband, Kilostream leased lines and private circuits were the best option for connecting business locations and computer networks.

However PSS eventually went the way of all X.25 networks and was overwhelmed by the Internet and more significantly the Internet's superior application suite and cost model.

Romanian Visit

'**Having** flown from Heathrow Airport to Bucharest Romania by TAROM, Romanian Airlines in the late 1970s for my second visit to secure a Monochrome Television Outside Broadcast Unit order, I was met on landing at the Airport by Hugh Lavington-Evans the Marconi European Representative based in Vienna. He had driven up that morning to Bucharest.

Over a period of seven days, various discussions had taken place with members of The Romanian TV organisation at the HQ of the Broadcasting Ministry. Eventually a Contract was signed but only after much bargaining and the handing over of several pairs of nylons and later a handheld Calculator.

On one occasion I was taken to the TV Studios of Romanian Television where I was shown the three Mark 1V Monochrome Cameras which I had sold to them sometime previously. I was then taken to a store where there were another six Marconi Mark 1V Cameras which on immediate inspection, were slightly different from those in the studio. I queried with the engineers where had these come from expecting them to say they had bought them second-hand from another organisation. To my horror, they advised these cameras were made in Poland to almost the exact design of our own UK made Cameras and that quite a number of these Cameras were being used by Eastern Bloc countries. This news certainly did not go down well when I advised our Technical staff as to what was happening. The Romanians were using Image Orthicon Tubes produced by EEV who were also surprised that they were being operating in the Polish made Mk 1V Cameras.

One incident which took place is worth recording. Hugh Lavington-Evans like myself, were Opera fans and we had heard that our Mk 1V Cameras were to be used at the Bucharest Opera House so we decided to pay a visit. The Opera, La Boehme was magnificent and we met the Romanian TV engineers who were well satisfied with the Marconi Cameras.

Having left the Opera House, we made our way back walking to our Hotel which was The Athene Palace in the main square of Bucharest approximately just over a mile away. We had not gone very far when we heard footsteps behind us, almost in tune with our own steps. After a few hundred yards, we turned round and saw a man with a long light coloured raincoat and wearing a Trilby hat about one hundred yards behind. We thought he was following us so we decided to nip up a side road but unfortunately we found ourselves in front of President Ceauşescu's Palace. We were challenged by the guards outside but did not understand what they said so we ran back through another road until we came to the Boulevard where our hotel was situated.

Turning round, we saw the same person in the light coloured raincoat following us so we ran quickly into the Athene Palace Hotel and advised the Concierge of our experience. The door opened and in came this chap with the light coloured raincoat. 'Monsieurs, I am very sorry' he announced in broken English ' I was at the Opera and saw you and followed you as best I could because I was lost and did not know my way back to the Hotel, I am very sorry to have concerned you'. Hugh and I were greatly relieved it was not the secret police and eventually this Frenchman who was also after business with the TV Authorities in Bucharest, stood us dinner and wine that evening.'

Peter A.T. Turrall MBE

THE RISE - AND FALL OF A DIVISIONAL MANAGER

'**Many** changes took place in Broadcasting Division over a period of years. The Sales and Contracts Departments were reorganised and new faces appeared in various managerial positions. I was appointed to direct the Overseas Sales with approximately twenty highly skilled sales people and a number of support staff and Secretaries.

The appointment necessitated moving offices so that all my staff were in one area. Responsibilities and areas of operation were set up after a short period and sales targets defined. Having got the operation underway, I decided it was time for an increase in my salary commensurate with the new position in which I found myself.

A new Divisional Manager had been appointed so I made arrangements to discuss my salary and possible increase with him. Having outlined what I had done since my appointment and the Sales targets I had set, I then asked for my salary to be increased in line with my new responsibilities. The Divisional Manager seated behind his desk on a swivel chair swung round quickly, the chair turned over and he ended up under his desk.

Obviously very embarrassed at the situation he found himself in, he agreed to raise my salary and I said I would not mention this incident outside his office. It was not long before the Divisional Manager took over other responsibilities within the Company. To this day I have not mentioned this incident to anybody else. I got the rise he got the fall.'

Peter A.T. Turrall MBE

By the mid-1990s the new Marconi plc considered itself to be a global communications and IT company, with 49,000 employees world-wide and sales in over 100 countries. It supplied advanced communications solutions and the key technologies and services for the Internet. Within Marconi plc were four business divisions: Communications, Services, Mobile, Systems and Capital.

The Company PR at the time wrote:

Marconi Communications.

A leading global provider of high performance communications solutions for public and private networks, Marconi Communications has a long history of innovation and technological breakthroughs. It is one of the world's rapidly growing broadband communications companies, helping to enhance the Internet with greater capacity and speed.

Marconi Services.

Marconi Services is focused on creating the communications solutions, architectures and networks of the future, offering customers the best in class services and products to meet their network needs. Its portfolio of services are marketed under a 'Plan, Build, Operate' banner. Its 'Plan' offer includes business and technical consultancy, network architecture and finance. 'Build' covers project management, civil works, network design, and integration. 'Operate' includes network support, network upgrade, network management and training.

Marconi Mobile.

Marconi plc believes that the market related to both mobility and security offers tremendous opportunities for growth and the Management of Marconi Mobile is gearing up to exploit these opportunities. Marconi Mobile starts from a strong base of state-of-the-art technologies: ATM, Radio Relay Communications, Mobile Communications, Secure Networks and Satellite Communications. The business is organised into two main areas covering Radio Mobile (Analog Mobile Radio, Digital Mobile Radio, Cellular Telephony and Air Traffic Communications) and Defence (Telecommunications, Defence Systems and Satellite Communications).

Marconi Systems.

Marconi Systems is the IT Division, employing more than 11,000 people worldwide and operating over 25 manufacturing and research facilities in eight countries. The division consists of three businesses - Commerce Systems; Data Systems and Medical Systems - which apply advanced electronic and information technology solutions to customers in more than 100 countries including hospitals, major oil companies, food and beverage companies, pharmaceutical and chemical manufacturers.

Marconi Capital.

Marconi Capital is a vehicle for the development of innovative high technology start-up and early stage investments to support Marconi's growth objectives for the future.

Everything seemed positive for the move into the Company's third century of operations.

It was a false dawn.

CHAPTER SEVENTEEN

Corporate Collapse

The Company that Marconi built was always said to offer a job for life. By the time I joined in 1983, there were often three generations of the same family working for the Company. But as the times changed, so did the Company. The dedication that the workforce gave to the Company for generations quickly dissolved.

Marconi's was once one of Britain's biggest and best known manufacturing concerns, but in 1946 the larger British Company English Electric acquired Marconi's. Marconi's complemented its other operations; especially heavy electrical engineering, aircraft and its railway traction business. In 1948 Marconi's was reorganised into four divisions:

- Communications
- Broadcasting
- Aeronautics
- Radar

These had expanded to 13 Manufacturing Divisions by 1965, when a further reorganisation took place. The Divisions were now placed into three groups:

- Telecommunications
- Electronics
- Components

By this time the Marconi's had facilities at New Street in Chelmsford, Great Baddow, Basildon, Billericay and Writtle, as well as in Wembley, Gateshead and Hackbridge. It also owned Marconi Instruments, Sanders Electronics and Marconi Italiana. In 1967 Marconi's took over Stratton and Company to form Eddystone Radio. During 1967/1968 the English Electric Company was subject to a takeover bid by the Plessey Company, but chose instead to accept an offer from GEC.

GEC, The General Electric Company Ltd, can trace its origins back to an electrical goods wholesaler established in London in 1886. In 1893, GEC had made the momentous decision to invest in lamp design and manufacture. The resulting Company became Osram in 1909, and it was to lead the way in lamp design, making GEC's fortune as the manufacturer of the most successful tungsten filament lamps in the industry. During World War One, GEC became a major player in the electrical industry with profits to match, the Company making radio sets, signalling lamps and arc lamp carbons.

During the 1920s GEC was involved in the creation of the National Grid, and in 1921 they opened purpose built headquarters in Magnet House, Kingsway, London WC2. Sir William Noble, a Director of GEC became one of the first Directors of the new BBC and GEC lent rooms in Magnet House to host the early days of the BBC.

During the Second World War, GEC was a major supplier to the military including the development of the cavity Magnetron for radar systems and the on-going production of lamps. The post-war years witnessed a slower pace to GEC's fortunes and expansion. Despite the demand for electrical consumer goods and large investments in heavy engineering and nuclear power, profits began to fall for the first time in the face of increasing competition, and internal disorganisation.

In 1961, GEC took over Radio and Allied Industries, and with it emerged the new power behind GEC, Arnold Weinstock, who became Managing Director in 1963. He moved the headquarters of the electrical giant from the historic Kingsway site to a modern building, at Number 1, Stanhope Gate. Weinstock embarked on a program which was to nationalise the whole electrical industry, but began with the interior rejuvenation of GEC. In a drive for efficiency, Weinstock made both cut-backs and mergers, injecting new growth and confidence in GEC, reflected in the profits and financial markets.

In 1967 the face of the British electrical industry was rewritten as GEC acquired Associated Electrical Industries (AEI), which encompassed famous Companies and brandnames such as Metropolitan-Vickers, BTH, Edison Swann, Siemens Bros., Hotpoint and W.T. Henley.

Despite this success, Lord Weinstock was not a man of vision. His risk-taking was really reserved for the racecourse. At work he could not have been more different. He would sit in his dowdy office, below pictures of his racehorses, behind a huge desk, a Dickensian figure peering at columns of figures which told him how and where the cash was flowing around the Company. He ruled the Company with a clique of old cronies and his ever-present confidante, Sara Morrison. I was once summoned with other Managers to see Lord Weinstock about some critical problem. We sat outside his office for nearly seven hours to be told that he no longer wished to see us. So we went home.

Good at numbers, he was hopeless as a leader of people, being quick to chide and slow to praise. He was a man who relied on seven financial ratios, such as the ratio of working capital to sales, which told him how healthy or unhealthy each one of the 180-odd GEC Companies was at any given time. At the first sign of trouble, he would pick up the telephone to bark at the hapless man in charge. He once fired a Manager over the phone in front of the astonished head of British Rail, Sir Peter Parker, after Parker had complained about GEC locomotives breaking down. Weinstock never visited his factories and rarely left the country, except for the opera in Salzburg, the races at Longchamps or Deauville, and visits to Israel. He appointed figureheads, such as a former Tory Cabinet Minister, Lord Prior, as Chairman, in order to attract clients such as the Ministry of Defence.

Margaret Thatcher had no time for Lord Weinstock, preferring more swashbuckling entrepreneurs. She thought he was yesterday's man, forever looking for state subsidies or protection. He also had a love-hate relationship with most of the brokers of the City of London: he loved to hate them, and they felt the same way about him.

GEC's share price languished, because he steadfastly refused to look for growth with computers, microchips, mobile telephones or consumer electronics. But Lord Weinstock saw that such ventures had lost millions for other European Companies, such as Siemens and Philips. He stuck to solid things such as defence electronics, turbine-generators and telephone exchanges, mostly sold to Governments in one guise or another. Spare cash was left in the bank rather than spent on takeovers. When rivals, in turn, tried to buy GEC in the late 1980s, he parked key businesses in turgid joint ventures with French, German and American Companies to render them impregnable.

Hence GEC shares were popular as a safe haven in recessions, because Lord Weinstock had a pile of several billion pounds in cash at the bank. His virtues of conservatism and caution, prized in the sluggish early 1980s and 1990s, were derided during booms. In 1968 GEC had merged with English Electric, incorporating Elliot Bros., Ruston and Hornsby, Stephenson, Hawthorn & Vulcan Foundry, Wilans and Robinson, Dick Kerr and Marconi's. The computer section of GEC, English Electric Leo Marconi (EELM), merged with Elliott Automation and International Computers and Tabulators (ICT) to form Marconi Elliott Computer Systems Limited (later renamed GEC Computers Limited) and International Computers Limited (ICL). In 1968 Marconi Space and Defence Systems and Marconi Underwater Systems were formed.

Marconi's continued as the primary defence subsidiary of GEC, and was renamed GEC-Marconi in 1987. GEC continued to expand, with the acquisition of Yarrow shipbuilders in 1974 and Avery in 1979. By this time, GEC had become Britain's largest private employer.

The late 1980s witnessed some major mergers within the electrical industry, with the creation of GPT by GEC and Plessey in 1986, and the acquisition of Plessey by GEC and Siemens the following year. An equal investment by GEC and Compagnie General D'Electricitie (CGE), formed the power generation and transport arm, GEC-ALSTHOM in 1989.

The movement towards electronics and modern technology, particularly in the defence sector, marked a change in direction away from the domestic electrical goods market. In 1990, GEC acquired parts of Ferranti and in 1995 acquired Vickers Shipbuilding and Engineering Ltd (VSEL).

Lord Weinstock, the GEC Managing Director retired in 1996 after leading GEC for nearly 30 years. He had helped it become the 18th largest business in Britain with a stock market value of £9.3 billion. Through a series of mergers and takeovers

GEC dominated Britain's engineering, electronics and telecommunications business.

In 1996, George Simpson took over as Managing Director of GEC, and with him came a new corporate management team. Simpson was an accountant and veteran of the car industry but he immediately set about dismantling the historic Company. A major reorganisation, aimed at focusing the Company on its 'key business strengths' was soon underway, involving the sale of Express Lifts, Satchwell Controls, AB Dick, the Wire and Cables Group, Marconi Instruments and GEC Plessey Semiconductors, and the planning of new alliances and acquisitions. In February 1998 the Company's head office moved into impressive and expensive new offices at Number One, Bruton Street, London W1.

In 1999 GEC underwent it most radical and major transformation. Marconi Electronic Systems (MES) was demerged and sold to Simpson's old friend Sir Richard Evans at British Aerospace for £7.2 billion, which then formed BAE Systems. GEC, realigning itself as a primarily Telecommunications Company following the MES sale, retained the Marconi brand, and renamed itself Marconi plc. BAE were granted limited rights to continue its use in existing partnerships, however by 2005 no BAE businesses used the Marconi name.

It was a move which managed to upset the French, the Germans and even Prime Minister Tony Blair in one fell swoop. Blair, who had made Simpson a life peer in 1997, wanted BAe to merge with DaimlerChrysler Aerospace of Germany. But most pundits agree Lord Simpson had got a very good price. Retaining the Marconi name, he brought in a new Finance Director, John Mayo, a thrusting young 'entrepreneur' with ambitions towards the telecoms sector.

Simpson and Mayo turned the Group's strategy on its head. After selling off all GEC's defence electronics and power generation divisions which had for some 40 years been its core businesses, he now decided that all efforts and focus would be on the new world of 'telecoms' which were at that time the stock market's darling.

The Company now went on a shopping spree, buying up expensive and over-priced telecoms companies in the US, plunging the Company into debt to jump upon the new hi-tech bandwagon. The problem, in hindsight, was that the wheels had already fallen off that bandwagon and nobody could put them back on again. In 1999 the American Internet Companies Reltec and Fore Systems were acquired by Marconi plc at the height of the 'dot-com' boom.

With the market's subsequent collapse the Marconi Corporation got into financial difficulties. As the Internet bubble burst, so did the Company's hopes. As Marconi's shares went into freefall, Lord Simpson and Mr Mayo were blamed for frittering away the £2bn cash pile they inherited from the old GEC and wasting £4bn on the over-priced acquisition of the two US Internet Equipment Companies. It had been a massively ill-judged gamble. Fortunes and tens of thousands of jobs were lost as a result. To survive the Company underwent a major restructuring. In a debt-for-equity swap shareholders were given just 0.5% of the new company, Marconi Corporation plc.

The name Marconi once stood for all that was great about British inventiveness. Now, it is the symbol of how badly things can go wrong. Guglielmo Marconi, who gave the Company its name, weathered many storms and financial crisis in his 40 years at the helm, must have been turning in his grave.

Monday, 3rd September 2001 was a dark day for Marconi's shareholders and employees. On that day, after months of reassurance that things would get better, Marconi's Chief Executive, Lord Simpson, was forced to admit that Marconi had lost hundreds of millions of pounds in just three months and that thousands of jobs would have to go.

On the day that Lord Weinstock died, the remains of his Company, once worth £35 billion was burdened with a £2.1 billion debt and valued at a mere £100 million. It was staring corporate oblivion and bankruptcy in the face. With sales falling further as debts rose, its share price was down to four pence from £12.50 at their peak. Despite his disastrous record, Simpson walked away from the carnage with a reported £1 million golden handshake.

The final blow came in May 2005 when Marconi missed out on a contract with its biggest customer, BT, worth a possible life-saving £10bn. Following months of speculation, in October 2005 the Marconi name and most of the assets were offered for sale to the Swedish firm Ericsson. The transaction was completed on 23rd January 2006, effective as per 1st January 2006. The Marconi name would still be used as a brand within Ericsson. The rest of the Company was renamed as Telent plc.

Today, the Marconi debacle has become notorious as one of the worst disasters in British corporate history.

The fate of the New Street factory was also sealed.

CHAPTER EIGHTEEN

The End of a Long Road

Long before the formation of Marconi's, Chelmsford became home to the United Kingdom's first electrical engineering works established by Rookes Evelyn Bell Crompton. Crompton was a leading authority of electrical engineering and was a pioneer of electric street lighting and electric traction motors within the UK. Crompton installed electric street lights around the town centre to celebrate the incorporation of the Borough of Chelmsford in 1888. Although this made Chelmsford one of the earliest towns to receive electric street lighting, the Council later decided to have it removed because gas was cheaper and the Council owned the Gasworks. Crompton supplied the traction motors for the first electric trains on Southend Pier. The Company also manufactured electrical switchgear, alternators and generators for many power stations in the UK and worldwide.

Crompton set up his original factory known as the 'Arc Works' in Queen Street in 1878. After a fire there in 1895, he built a huge new electrical engineering factory also called the 'Arc Works' in Writtle Road. The Firm was called Crompton and Co. and in 1927 became Crompton Parkinson after Colonel Crompton formed a business partnership with fellow British electrical engineer Frank Parkinson. During World War II, the factory was frequently targeted by the Luftwaffe. In 1968 Crompton Parkinson Ltd was downsized and operations moved elsewhere after a takeover by Hawker Siddeley. The site was taken over Marconi's and became the base for the newly formed Marconi Radar Systems Ltd when the radar department moved out of New Street.

After years of decline, the Marconi Radar factory finally closed in 1992 and the site was demolished a few years later apart from the frontage on Writtle Road. A housing development called 'The Village' now occupies the site with road names such as *Rookes* Crescent, *Evelyn* Place, *Crompton* Street and *Parkinson* Drive as tributes to the former occupant. The destruction of the Marconi Radar site carried on to include the Marconi Waterhouse Lane site that was completely redeveloped as a business park.

Next to Marconi's New Street Works in Chelmsford was the United Kingdom's first ball bearing factory. It was established at New Street and Rectory Lane in Chelmsford in 1898 by cousins Geoffrey and Charles Barrett. American ball bearing machine manufacturer Ernst Gustav Hoffmann (from whom the Company took its name) supplied the patents for the early ball turning machines. The Hoffmann Manufacturing Company rapidly expanded and soon achieved worldwide fame for their precision-made bearings. Hoffmann bearings were later used in the first transatlantic flights and extensively on machinery during World War I. For many years it was Chelmsford's main employer with even more staff than nearby Marconi's.

The firm became R.H.P.(Ransome Hoffmann and Pollard) after amalgamation with the Ransome and Pollard bearing manufacturing companies in 1969. The factory that once employed 7,500 employees over 50 acres was wound down during the 1980s and finally closed for good on 23rd December 1989. The Company assets and name were absorbed into the Japanese NSK Ltd Bearing Company in early 1990 trading as NSK-RHP Ltd at its UK base in Newark on Trent with the historic R.H.P name finally disappearing in 2001. Most of the factory was demolished during the summer of 1990 and the site is now occupied by the Rivermead Campus of the Anglia Ruskin University.

The Hall Street factory building served for many years as the Mid Essex Divisional Offices of the Essex Water company and is still private property. The world's first wireless factory has survived with the exterior more or less unchanged. The Marconi work's sign and the ivy have long since gone but the building can be considered to be the oldest electronic factory in the world. However, it was reported in December 2009 that Anglian Water, the owner and current occupier of the site would shortly be vacating the site. It will be sold for housing and conversion of the existing building into flats. In October 2011 the site stood empty and at the time of writing its future is unclear.

Hall Street Factory, October 2011

The Marconi College at Arbour Lane has also gone. The bulldozers moved in during the summer 2002 and demolished the whole site except for the front building known as 'Telford Lodge'. The site has now been developed for housing.

By the turn of the 21st Century, the original Marconi New Street Works housed what was left of the Military Communications Division of Marconi Communications, now known as Selex, a branch of the Italian electronics firm Finmeccannica. They were to be the last occupants of the site.

By 2006 Chelmsford Borough Council had already recognised that the site had problems. They noted in their 'Employment Land Review' for 2006 that:

> 'The level of activity on the site has been in gradual decline with over half of the available floor space now currently unused. Selex

Communications intends to cease use of the site and transfer existing operations elsewhere in the local area making the site operationally redundant. The accommodation is generally outdated and already requires extensive repairs and maintenance. The accommodation on the site is not of the standard required for a modern high-technology Company, and the extent of floor space is excessive in relation to Selex Communications' current needs.'

Perhaps the most significant statement was the cost of the site as the rateable value for the entire site as at 1st April 2005 was £1.575 million. In 2008 Selex Communications moved out, ending nearly 100 years of communications industry and radio history on the site.

Ongoing changes meant that with the exception of the new (ex-Marconi radar) Eastwood House that belongs to BAE Systems, the remainder of the original Marconi site was sold to developers, Messrs. Ashwells of Cambridge. The Company originally retained Mr Ian Parkin to put together a proposal to tell the story of Marconi and the Company on the redeveloped site. It was planned that most of the New Street site would be demolished and the new buildings would be mainly residential and commercial with the site being essentially restricted to pedestrians. Across the road the Drivers Yard car park in Victoria Road was closed as was 'The Laurels' house, now demolished and developed as shops and an apartment block.

On the main New Street building only the 1912 frontage is protected by listed status. It was hoped that the power house, water tower and the semi-detached cottages on the corner of Marconi Road would be kept. It was suggested that the new buildings, footpaths and squares would have names, particularly one called Building 46, and that the names should reflect both events in Marconi's history and the names of significant people in the history of radio. There was even a plan to provide a Marconi Heritage centre in the front building.

> **'First** of all let me advise you of the situation as it was in Chelmsford at the early part of October 2005. The Planning Design team of Chelmsford Borough Council had drawn up plans for the ultimate redevelopment of the Marconi New Street site. This included the removal of Building 720

and also Marconi House and the large factory area. A road would be inserted to give access to the Anglia Ruskin University at the bottom end of New Street. The front building would not be altered (I managed to get a preservation order on it long before I retired from the Company) but extensions would be made at the rear of this part of the building and where the old factory stood, both housing, light industrial units and a car park would be included. Even part of the railway embankment was required in the overall plan.

However, on 14th October, a Press Release was issued by Selex, the owners of the site, to state that in 2007, the people employed at New Street (some 390) will be moving to an old Marconi site at Christopher Martin Road, Basildon (part of the old Avionics Group). This came as a great surprise to Chelmsford Borough Council who, up until this time, had been in discussion with the management of Selex. The next we heard was that representatives of Chelmsford Borough Council were meeting the chosen developers of the site, with a view to try and get the developers to give the front building to the council. We do not at the time of writing, know the answer to this question. Our local MP Simon Burns was involved in the discussion as the Council probably wanted to site the Museum at this place. The front building includes the original office of Gugliemo Marconi which I had the privilege of occupying for the last twelve years of my service with Marconi Communication Systems.

We understand that the requirements of the site developers are to build houses on this 30 acre area. However, a rumour exists that Marconi House, the four storey building, might be turned into flats. The new Eastwood House building currently occupied by BAE Systems, is not affected in these negotiations.'

Peter Turrall MBE

On 21st October 2008 with application number 08/00450/FUL Ashwell (Chelmsford) Ltd and their planning consultants Bidwells (Chelmsford) secured a resolution to grant planning permission to develop the former Marconi Works and Railway Station in New Street, Chelmsford, Essex. The proposal was for mixed use redevelopment and conversion of some existing buildings for residential, office, retail, financial and professional services, drinking establishments, hot food takeaway and non-residential institutions. It also included a new train station forecourt, access arrangements, alterations to Townfield Street multi-storey car park, cycle storage building, diversion of public footpath and associated access parking and landscaping.

A parallel application for listed building consent was also approved to provide for demolition of the factory and the retention and refurbishment of the key historic buildings. The redevelopment would provide a new urban neighbourhood comprising 715 new homes, a 120 bed hotel, 30,000m² of commercial floorspace and a new arrival space for Chelmsford's rail station.

Designed by Rogers Stirk Harbour and Partners, the scheme represented: 'A modern urban quarter which maximises the site's development potential yet at the same time respects the key historic elements.'

Lee Melin, Partner at Bidwells said at the time: 'The Marconi factory was synonymous with technological innovation and progress: the RSHP masterplan is a fitting celebration of this legacy.'

Development work was planned to begin in 2010, but in December 2009 it was reported that the owners had gone into administration. When Ashwell Property Group collapsed at the end of 2009, it took with it one of Chelmsford's largest and most ambitious regeneration plans worth £40 million.

By June 2010 it seemed that Chelmsford Council was being forced into taking legal action to safeguard the New Street site after an investigation by BBC Essex uncovered extensive vandalism and trespass at the site.

At a meeting of Chelmsford Council's cabinet, Neil Gulliver, the councillor responsible for planning and building control, said an enforcement notice would be issued to the receivers of the site for a schedule of work to be carried out to safeguard its future until any development took place. He said: 'It must be kept secure and maintained in an appropriate manner.'

'We had hoped to reach an amicable agreement with the receivers, but this was not possible so the notice gives them about eight weeks to carry out the work.'

Chelmsford MP Simon Burns also criticised the state of the building.

He said: 'It is both a disgrace and appalling that the receivers, who currently own the site, have not taken efforts to protect this historic site, which has listed building status.'

'This cavalier attitude to the property they own has led to anyone being able to breach its boundaries, cause damage and commit acts of vandalism, which is destroying the interior of the building.'

'It would seem from the evidence that squatters have taken up residence and there is evidence of drug abuse, the consumption of alcohol and graffiti as well as wanton damage to the interior.

'This will cause grave distress to the many people in the town who worked for the Marconi companies and to those who value our history and heritage.'

A spokesman for receivers Zolfo Cooper said: 'Prior to our appointment as LPA Receivers of the former Marconi site, in November 2009, a substantial amount of vandalism and theft had occurred on the site, resulting in it becoming unfit for tenant occupation and making it more difficult to secure for an extended period of time. We have continued to operate regular security patrols on a daily basis, in line with the arrangements made prior to our appointment, while we have considered the options available for the vacant site.'

'We recognise the Marconi building in Chelmsford has continued to be the target of trespassers and as a consequence we have instigated a thorough independent security review in recent weeks. As a result of this review, started before the BBC's filming, security arrangements have been tightened further, including the repair of perimeter fencing and gates which were damaged by trespassers intent on gaining access to the site.

'In addition, we have also introduced a motion detection system which provides round-the-clock supervision of the site.'

The site is now securely boarded up but it is suffering from severe neglect including previous vandalism that has in part damaged the roof allowing water to enter. All around the site weeds have taken a significant foothold.

By August 2011 it seemed that a number of construction firms were locked in a bidding war for the disused Marconi site. In August Neil Gulliver revealed he has been shown two potential plans and that it was just a matter of bidders negotiating the right price with the administrative receiver, Zolfo Cooper.

He said: 'One project is for about 200 houses, the other is very much along the lines originally proposed by Ashwell Land with about 900 homes in flats, townhouses and some commercial redevelopment.'

It is possible that the site could be bought by a reincarnation of the company that lost the site when it went into administration in 2008. Brookgate, the property company that emerged from the collapse of Ashwell Property Group, is thought to be one of developers bidding for the Marconi works and factory in Chelmsford. The bidding process ended in July 2011.

Ashwell had bought the site in 2008 for £25m. Zolfo Cooper put the site into receivership when Ashwell went into administration in 2009. The Marconi site on New Street has been vacant since 2008 and is now worth around £10m.

It seems that Zolfo Cooper has been slow to decide on a preferred bidder, despite pressure from local authorities. A spokesman from the council said: 'We have pressed Zolfo Cooper to find a way forward in relation to developing the site. There are three potential owners, but there is no clear favourite at this point.'

Councillor Neil Gulliver said: 'The development of Marconi site on New Street has very much been stop-start, but I do not think that is atypical of what has been happening across the country. Getting this site developed is really subject to the economy recovering.'

At the time of writing, the future of the historic Marconi New Street site is still unclear.

Only time will tell what will happen now.

The Marconi New Street Works was the world's first factory to be built specifically for the manufacture of wireless components and equipment.

It can rightfully claim to be the birthplace of the modern electronic age.

However I fear that like so many other sites of industrial archaeology that shaped our modern age, including the 2MT Writtle site and 2LO's building on the Strand in London , instead of shaping and defining the rich heritage of this country for future generations, it will simply be lost forever.

CHAPTER NINETEEN

After-words in 2012

One Hundred Years

The development of wireless revolutionised warfare and safety at sea. It allowed nation to speak unto nation and even allowed us to communicate with men walking on the moon and space craft as they leave our solar system.

Wireless and then the television also completely changed the way ordinary people lived their lives and even changed the very layout of their homes. No longer was the fire and hearth the focus of attention in the main room as it had been for over 400 years. Now the wireless and later of course the television dominated their lives and for the first time people no longer had to entertain themselves.

With wireless the whole world quickly came to expect a rapid and accurate news service, a varied and comprehensive entertainment service and a source of education. Suddenly there was a need for schools and training colleges to teach all aspects the new science. It also brought about the development of new industries and factories to manufacture and maintain the equipment, creating employment and helping the formation of large towns. As radio developed, new international governing bodies had to be set up to regulate frequency allocation and transmitter output powers. Wireless also changed the science of criminology and law enforcement. The world became a far smaller place with nowhere to hide.

Marconi and his Company built the wireless age. The full list of inventions, projects and patents would fill many volumes, but amongst the great achievements that changed the world we live in are:

Wireless tuning and the diode valve in 1904.

Transatlantic wireless communication and the formation of Marconi Marine.

Radio direction finding.

Radio telephony.

Airborne wireless.

High Frequency tuned broadcasting.

The formation of the British Broadcasting Company that later became the independent BBC.

Formation of the Marconi Wireless Telegraph Company of America (later RCA).

Short wave beam broadcasting.

Radar.

Television.

Modern avionics.

Much of it was designed, developed, built, tested, packed and shipped within the walls of what we called the Marconi New Street Works. What was done there changed the face of British society and indeed the world forever.

But of course it has all gone now.

New Street 2011

GUGLIELMO
MARCONI
1874-1937

THE FATHER OF WIRELESS
From this site was
transmitted Britain's first
official radio broadcast
by Dame Nellie Melba
15th June 1920

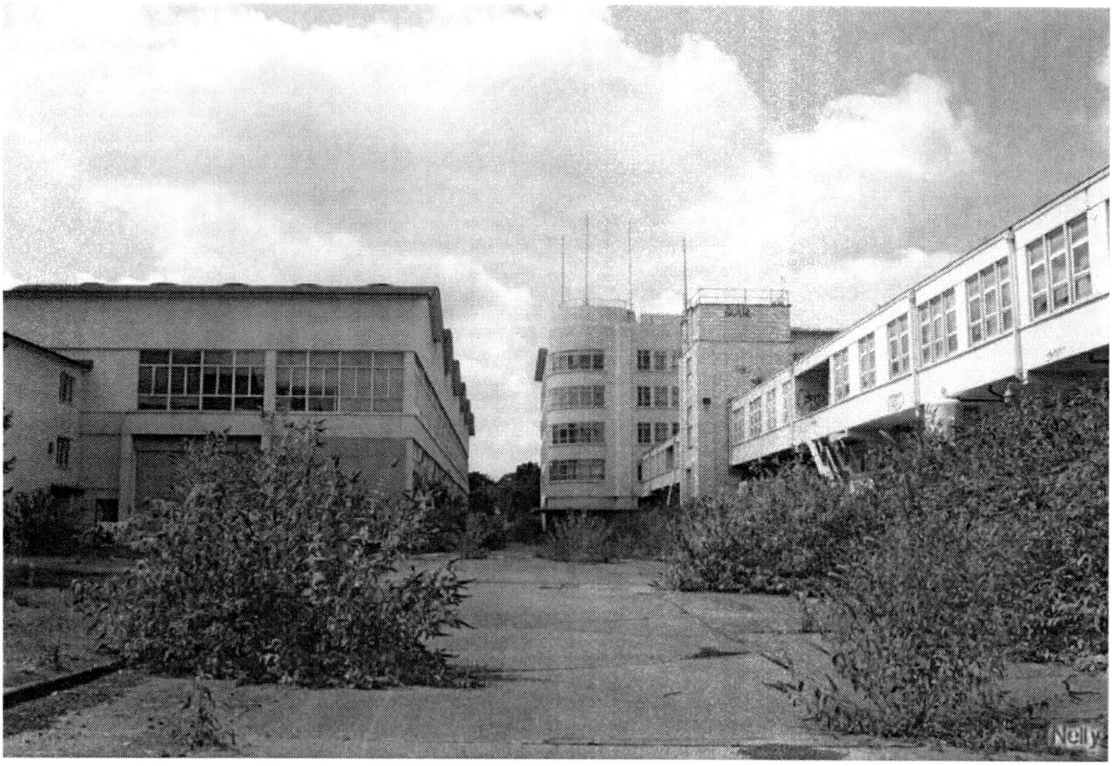

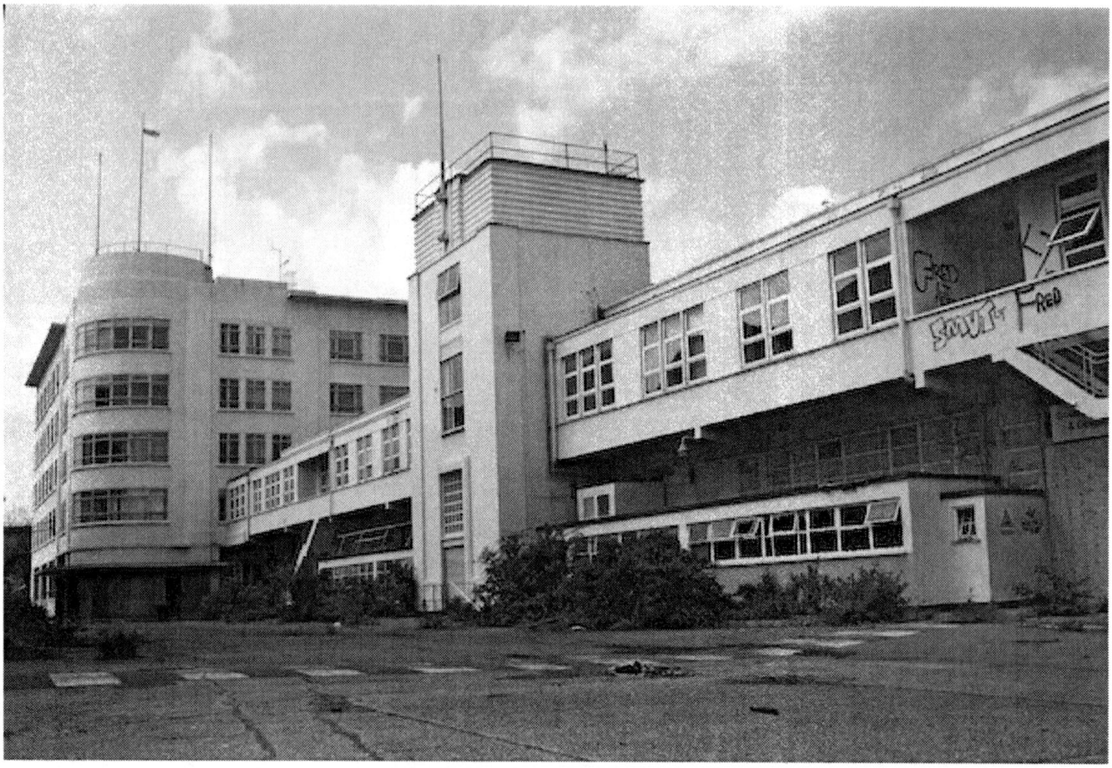

Overgrown, vandalised and abandoned - Marconi New Steet 2011

The scrapheap beckons - the ignominious end of several eras

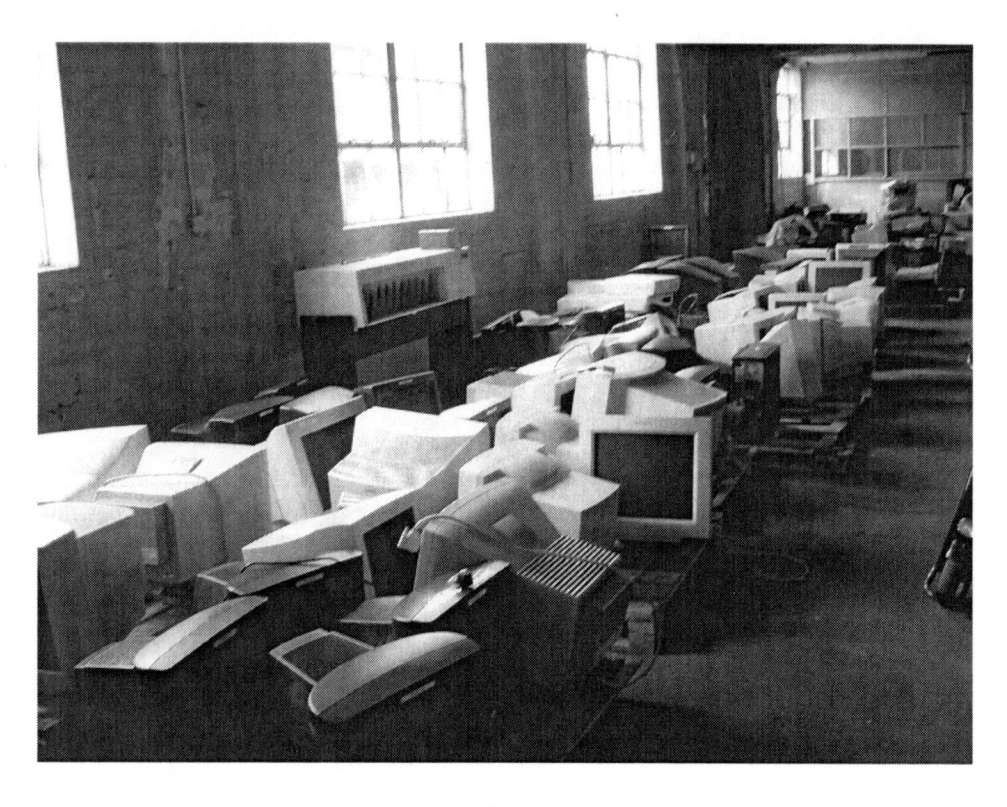

The last person who left the works did indeed turn out all the lights......

The grand main entrance to New Street in better times. Marconi sometimes used the office on the righthand side of the landing

The foyer to Marconi House as I remember it

The main yard looking toward front gate

The roof of building 720

The works canteen, ground floor building 720

After more than seventy years the factory still bears the scars of wartime explosion and shrapnel is still buried in the wall to this day

The goods yard, water tower and wartime camouflage

Abandoned and at the mercy of the elements
New Street, July 2011

The top floor of bld 720, my old stamping ground - once the offices of CSU

Loading bay, Marconi Goods yard

When I started this project I asked a simple question – '*Do you remember?*'

Again I would like to thank the very many of you who did. I asked the question because I think that it is important that someone took the time to write down what people remember about the once great Works, while there is still time to remember. Sadly that is why the first thirty years or more of the works early career had to be documented as a historic text, simply because there is no one left who remembers.

Also if we don't record and learn the lessons of the past then perhaps we are bound to repeat them. The eyewitness accounts of people desperately searching for family members in bomb damaged factories and houses, while Christmas decorations lay in the road are perhaps the saddest and most haunting images I have ever written about.

Perhaps not for today, but in another hundred years' time there will be no one left who worked there and the site will be a distant memory or just another faded blue plaque on a nearby building, I am sure that someone will be interested to know something of what happened there. Perhaps, as I have done, they will be researching the origins of everything that built their modern world.

So to that person, sometime in the future, may I take this opportunity to wish you well across the years and hope that you have enjoyed the story. If you get the chance, please tell it to others and tell it often.

Although we perhaps never realised it at the time, Marconi's New Street Works was a great place to go to work every day. To my future reader also please understand that amazing thing happened there – I recently wrote in another place about another time that the development of wireless was a kind of magic and New Street was part of that magic as well.

As for the New Street Works building in its Centenary Year, well I fear that the story is almost over. Time has all but stopped there now apart from the insidious influx of weather, nature and vandals. It will never again spawn a technical revolution or help its country win a war. The last person who left, turned the lights out forever and walked into the night.

But please also realise that in these pages I have only scratched the surface.

Over a hundred years there are countless thousands of forgotten names and faces, forgotten projects and no doubt vital deadlines. We all in our time as Marconi men and women weathered a thousand crises. We also shared much laughter, love and a fair proportion of tears. There is sadness too for the 17 people who went to work one evening and didn't go home.

I remember walking though the Marconi Writtle site just before it was demolished and although only 40 years old, it too was a testament to ingenuity and invention and everything that went on within its walls.

Recent photographs of the New Street Works show a haunting interior, condemned by age and wearied by neglect. Laying around there are still faded and poignant glimpses of the past. A discarded fire extinguisher, forgotten punch card machine, an old telephone directory, decaying manuals and many warning signs, now mute and unheeded. If only the walls could talk, what a tale they could tell.

I am also sure that there must be ghosts there. Every old building with such a amazing history has them - they haunt the corridors and machine shops, huddling together and hiding away as you walk down New Street and glance up at the windows.

I hope that they and you enjoyed this story.

Tim Wander
January, 2012

APPENDIX ONE

MARCONI'S NEW STREET WORKS
A HUNDRED YEARS OF CHANGE

1912

The Marconi New Street Works throughout its entire lifetime was subject to continual changes. From the days it first opened its doors until practically the last weeks, its layout and organisation was often driven by projects, equipment and the space needed for large machine tools or manufacturing lines. Internal partitions and temporary huts sprung up and were swept away on an 'as required' basis. The maps here show the Works at various phase of its development.

In 1913 another landmark appeared over Chelmsford at the Marconi Works with the erection of two 450 foot wireless masts. The masts and antenna can be seen in the plan of the Works for 1921 below. Marconi always made use of army huts to house his various developments. These can be seen around the factory building, in the courtyard and to house the transmitter at the base of the antenna feed. Sprinklers had been added to the pond to prevent stagnation.

On the other side of New Street, opposite the Factory were the Railway Goods Yard and the stables. As the goods trains pulled up, various horse and carts would stop here to collect their goods. Key amongst them would be the different coal merchants, including Charringtons and Moys horse and carts, ready to load up and deliver to the customers.

Goods destined for the Marconi Works were shunted onto the private siding and across a single railway track that crossed New Street connecting the main factory via the 2nd gate nearest to Marconi Road to the Railway Goods Yard and hence to the main line. Traffic and pedestrians would have to stand and wait as the carriages trundled across the main road. The factory had its own loading platform along the North side. The platform still exists pending the whole site's future demolition. The track was believed to have still been visible in 1969 and its impression until the mid-1980s.

MARCONI'S NEW STREET WORKS

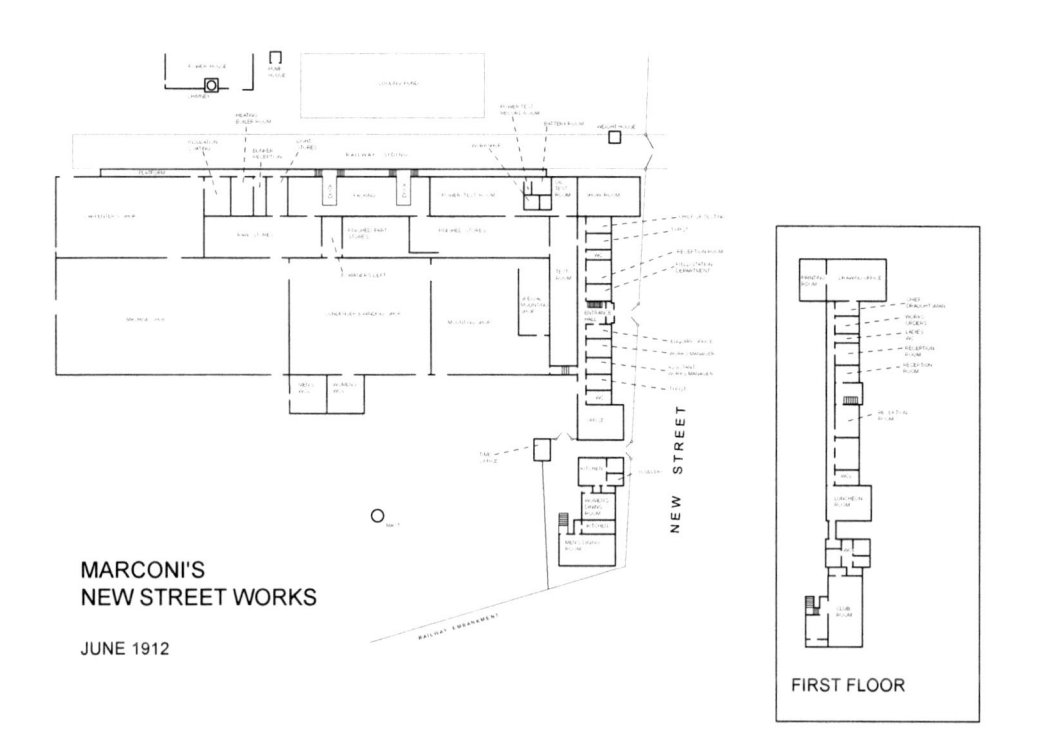

JUNE 1912

FIRST FLOOR

Marconi New Street Works, Front View, c1915

MARCONI'S NEW STREET WORKS in 1921

Extensions to the 1912 factory

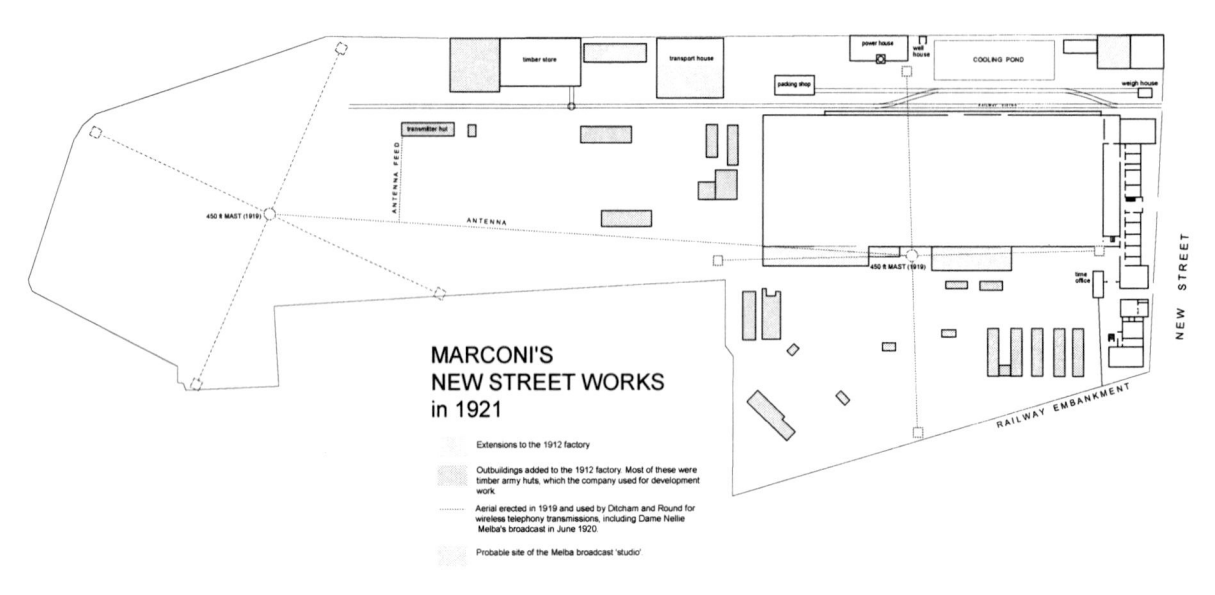

Outbuildings added to the 1912 factory. Most of these were timber army huts, which the company used for development work.

Aerial erected in 1919 and used by Ditcham and Round for wireless telephony transmissions, including Dame Nellie Melba's broadcast in June 1920.

Probable site of the Melba broadcast 'studio'

Marconi's New Street Works, Aerial Views, c1920

Marconi New Street Works, Aerial Views, c1920

1927

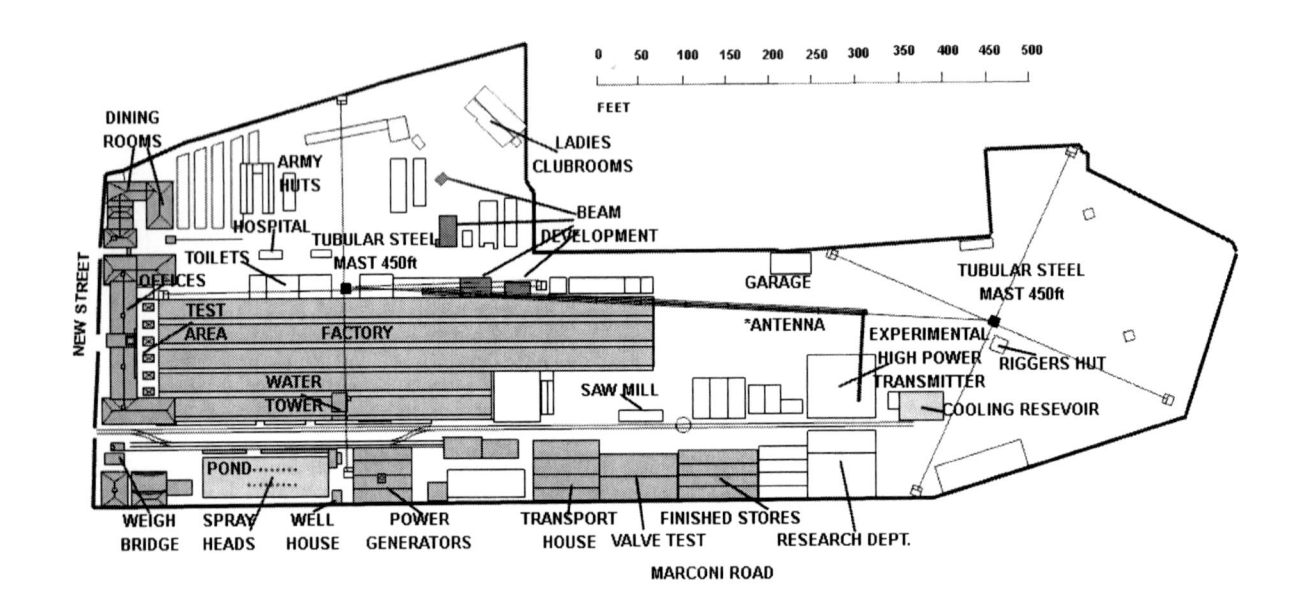

By 1927 the development of the Marconi short-wave 'Beam' system and equipment was underway with ambitious plans to supply the 'Imperial Chain'. More factory space was required so the factory floor was extended with four areas dedicated to development of this system. The design of this extension matched the original 1912 plans for the factory. A second borehole was sunk at the western end to supply a cooling reservoir for the high power transmitters. The Marconi Research and Valve test department was started although it would later break away from New Street to other factories in the Chelmsford area.

Marconi New Street Works, Frontage, c1930

Marconi New Street Works, main yard, c1930

1936

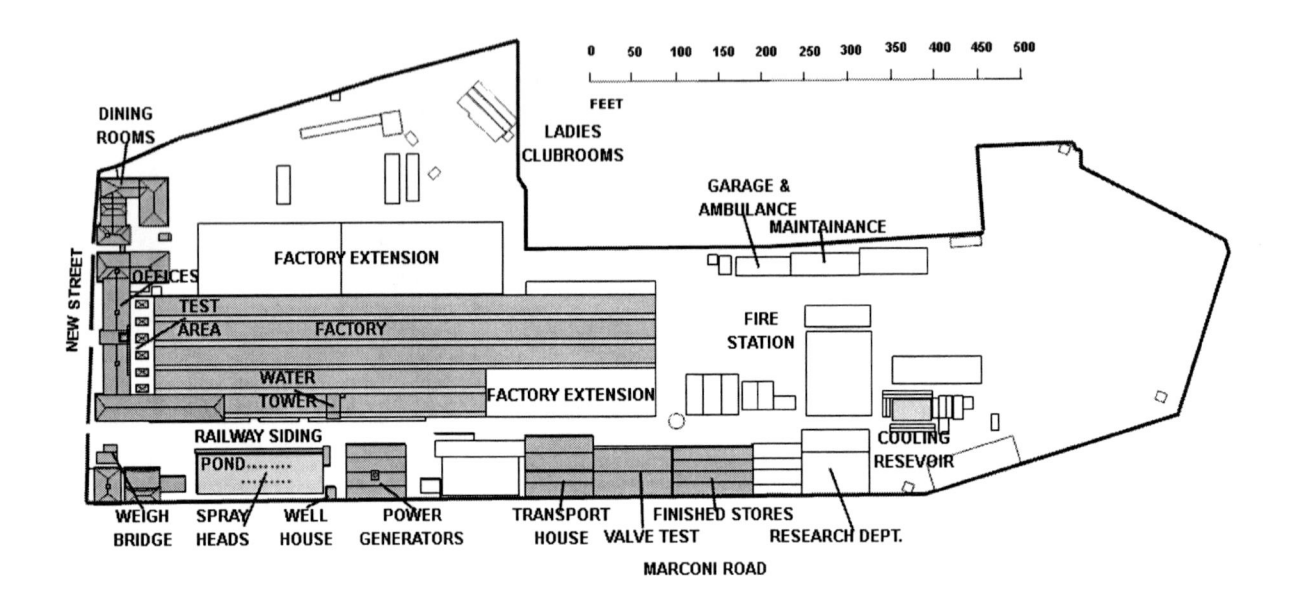

By 1936 the two 450 foot masts have been removed. The test area had becoming too small and the factory is once more extended. This time little care, probably due to recession and possible fear of impending conflict it seems that little car or thought is taken over the aesthetics. The new sections are plain square brick sections which cover up many of the original arches facing the main yard. The whole of Marconi's undergoes reorganisation when the Marconi Research Centre at Great Baddow opens. Then called the Marconi Research Laboratory, it was originally built by to bring together their various radio, television and telephony research teams in a single location.

1938

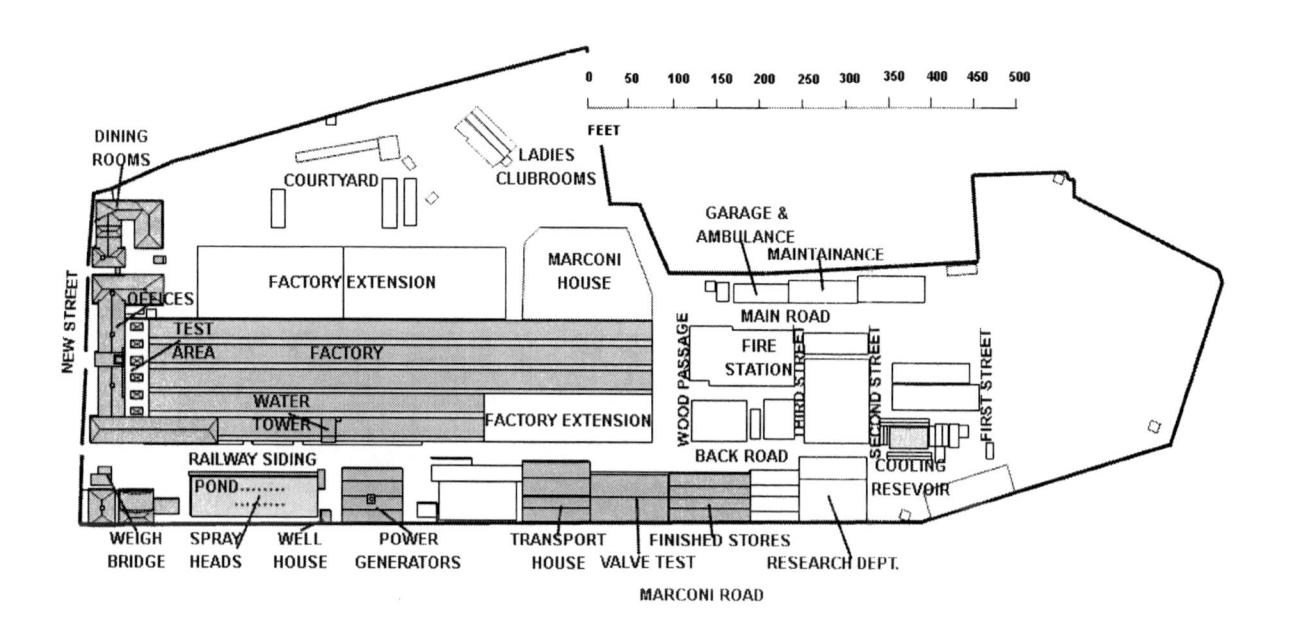

1938

In 1937 some of the adjoining land was purchased for the construction of Marconi House to house the main company offices is undertaken. The Managing Director had an office on the fourth floor with en-suite shower and toilet.

1941

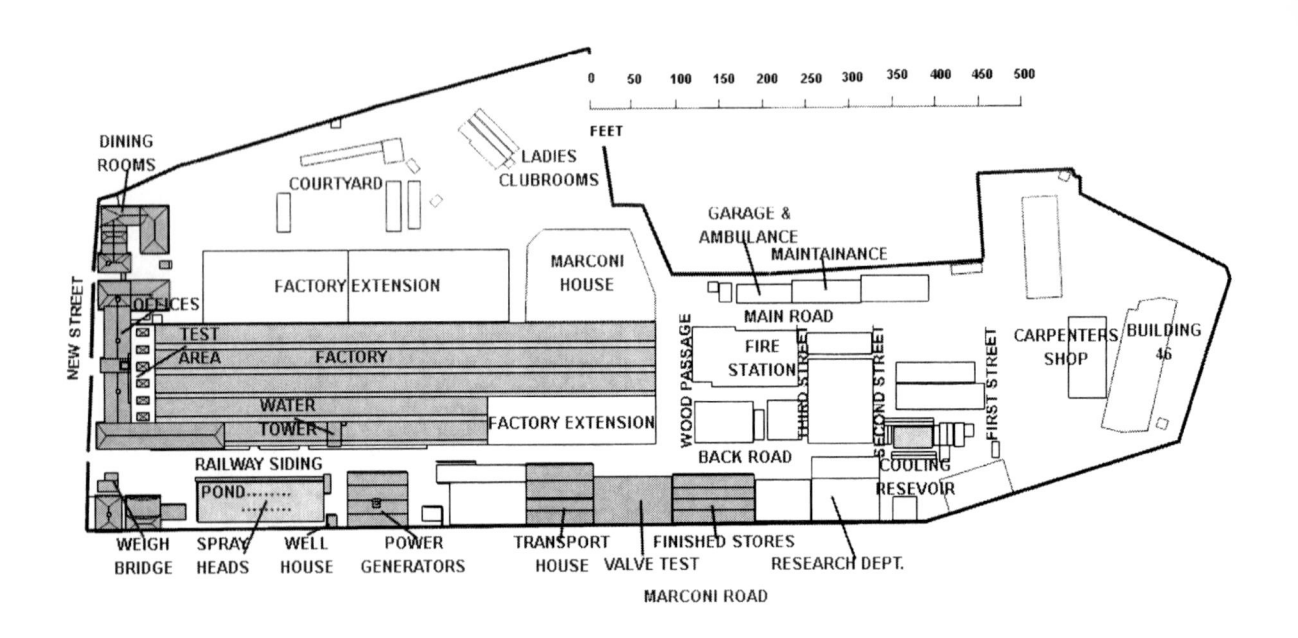

1941

In 1941 Building 46 is built for the development and manufacture of high power broadcast transmitters.

New Street Works, 1943 Luftwaffe Aerial Target Photograph

1949

The girls pavilion is removed and building 720 is built. Building 720 was unique at the time in having the largest unsupported roof span in the country.

In 1960s, Building 720 used the whole 2nd Floor for catering with three adjacent canteens. Furthest from New Street was a Works Canteen (long queues and collect, two/three serving hatches), adjacent was a waitress canteen (same menu) and then a third 'semi-important' silver service one (nearest to New Street) that had another menu for more senior staff and visitors. Another separate Senior Directors and Managers Executive Restaurant was in the main building adjacent to New Street.

Also in the 1960s, the 1st Floor of building 720, was the Apprentice Training Centre (ATC) Admin offices and outside 30 clerical/secretarial trainees worked at their typewriters. The annual apprentice/graduate/clerical intake in 1969 filled the MASC Dance Floor on the first day induction. On the ground floor, production of large system racks occurred and, down a short ramp, the infamous 'PIT' where all apprentices learnt in year one all the basics of the machine shop's tools.

Marconi New Street Works, Building 720 under construction, c1949

Marconi New Street Works, Main Yard c1950

Aerial Photographs, c1955

Building 46

The air raid shelter under building 46 – used for many years as a store and dumping ground for old plans and spares. When building 46 was demolished the shelter was filled with concrete – much of the unwanted assorted spares, plans and ephemera was left down there and encased - perhaps for future industrial archaeologists.

MARCONI'S NEW STREET WORKS 1969

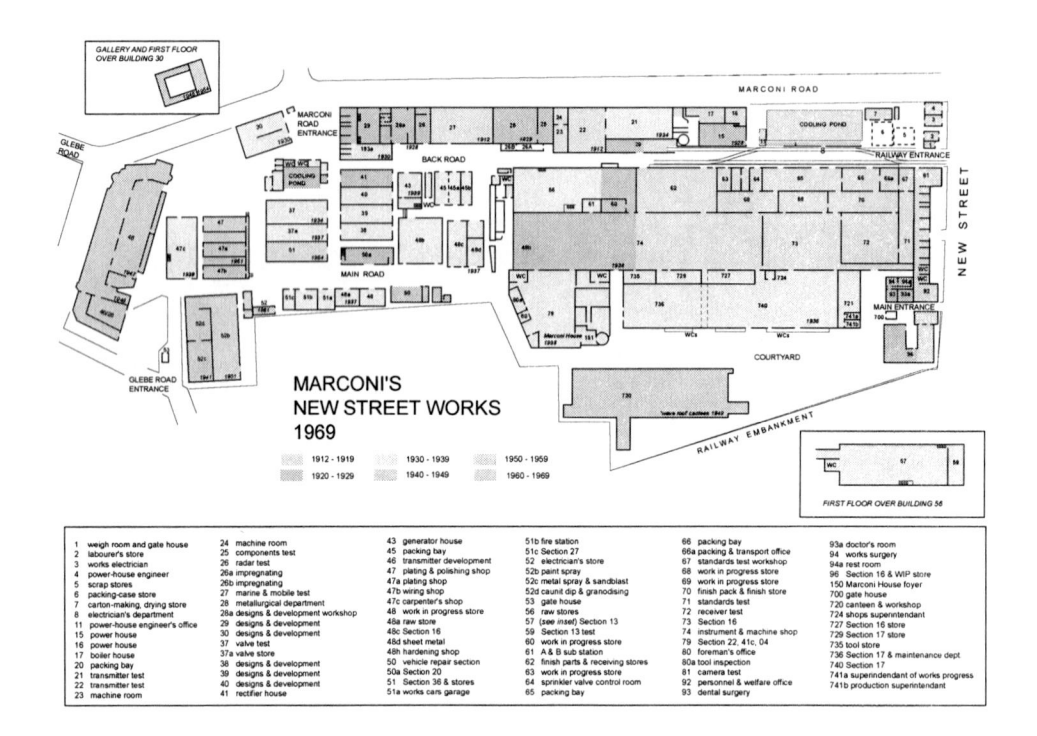

1912 - 1919	1930 - 1939	1950 - 1959
1920 - 1929	1940 - 1949	1960 - 1969

1. weigh room and gate house
2. labourer's store
3. works electrician
4. power-house engineer
5. scrap stores
6. packing-case store
7. carton-making, drying store
8. electrician's department
11. power-house engineer's office
15. power house
16. power house
17. boiler house
20. packing bay
21. transmitter test
22. transmitter test
23. machine shop
24. machine room
25. components test
26. radar test
26a impregnating
26b impregnating
27. marine & mobile test
28. metallurgical department
28a designs & development workshop
29. designs & development
30. designs & development
37. valve test
37a valve store
38. designs & development
39. designs & development
40. designs & development
41. rectifier house
43. generator house
45. packing bay
46. transmitter development
47. plating & polishing shop
47a plating shop
47b wiring shop
47c carpenter's shop
48. work in progress store
48a raw store
48c Section 16
48d sheet metal
48h hardening shop
50. vehicle repair section
50a Section 20
51. Section 36 & stores
51a works cars garage
51b fire station
51c Section 27
52. electrician's store
52b paint spray
52c metal spray & sandblast
52d caunt dip & granodising
53. gate house
56. raw stores
57. (see inset) Section 13
59. Section 13 test
60. work in progress store
61. A & B sub station
62. finish parts & receiving stores
63. work in progress store
64. sprinkler valve control room
65. packing bay
66. packing bay
66a packing & transport office
67. standards test workshop
68. work in progress store
69. work in progress store
70. finish pack & finish store
71. standards test
72. receiver test
73. Section 16
74. instrument & machine shop
79. Section 22, 41c, 04
80. foreman's office
80a tool inspection
81. camera test
92. personnel & welfare office
93. dental surgery
93a doctor's room
94. works surgery
94a rest room
96. Section 16 & WIP store
150. Marconi House foyer
700. gate house
720. canteen & workshop
724. shops superintendant
727. Section 16 store
729. Section 17 store
735. tool store
736. Section 17 & maintenance dept
740. Section 17
741a superintendant of works progress
741b production superintendant

New Street Works, looking toward Building 46 from Building 720

Marconi New Street Works, Goods Yard and Marconi House, 1980s

1980

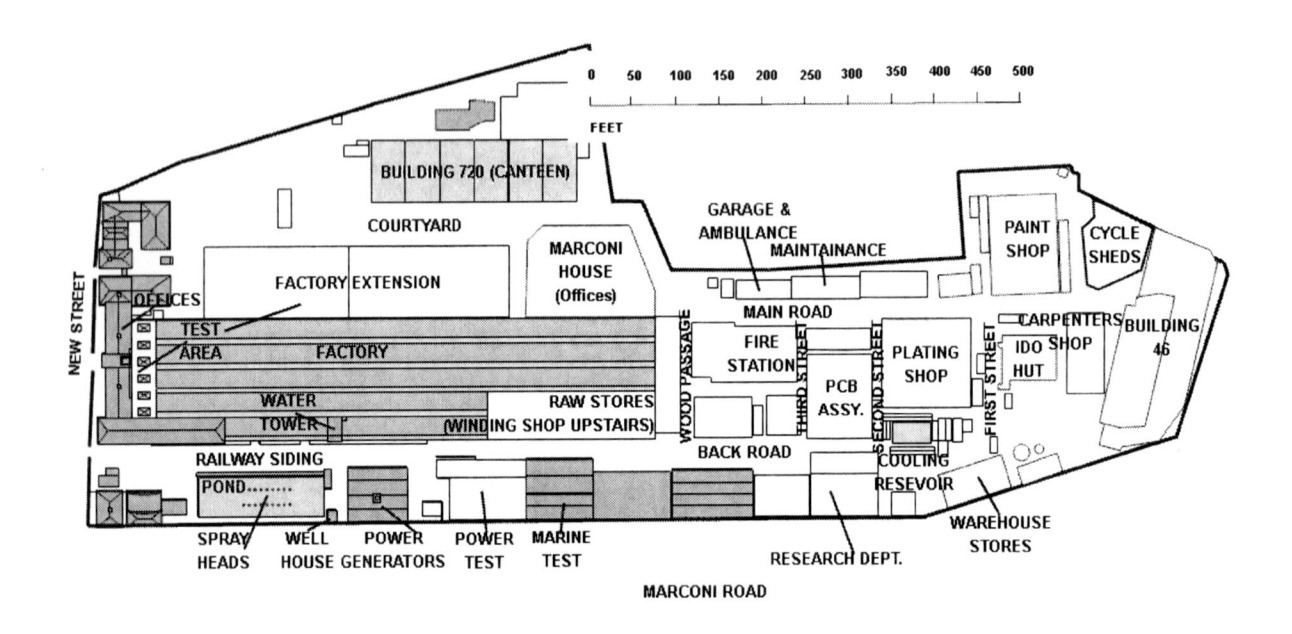

1980

The drawing shows the factory layout in 1980, at the peak of the Company's expansion. Building 46 has been extended with a second floor above. The IDO hut is located near to the Carpenters Shop. Printed Circuit Boards are now commonplace and a dedicated PCB section lies between the first and second streets.

On the Goods Yard side of New Street (opposite the factory) but on the Victoria Road side of the railway bridge was a large site known as 'Drivers Yard'. It extended from the railway line to Victoria Road and had significant depth.

Access was from both New Street and Victoria Road. This was the base for the Company Chauffeurs and their cars. Company fleet cars were based here and there was a Company Petrol Pump. Buildings adjacent to Victoria Road were originally used for a Social Club until the MASC in Beehive Lane was built circa 1960. The Marconi Apprentice Association (MAA) used some buildings in the late 60s and the Marconi fire engine was stored there for many years

The site also housed the Company 'Print Works'. At the farthest end, was 'Reclaim Stores' or 'Disposals'. This provided employees with surplus components, wire and test gear very cheaply - most small things were 6d, test gear was more, perhaps a pound or so. Employees developed their skills building their own radios, TV's and stereos and the Company benefited from their self-interest and skill development in their own time; losing out perhaps with subsequent problem solving during the day. An Electronics Component Shop further up New Street, towards the Cathedral, and whose prices were slightly higher could hardly have appreciated the competition.

For a time the GEC-Marconi Company Publicity Department archivist, my old friend Roy Rodwell were housed in the 'Laurels' – they also seemed to keep busy printing some form of Marine Reports.

In 1987 the historic Marconi Writtle site closed and the staff and laboratories transferred to the Marconi New Street site. The top floor of building 720 was redeveloped as the Computer Systems Unit.

Marconi New Street Works
Demolition for Eastwood House Construction, c1992

2000

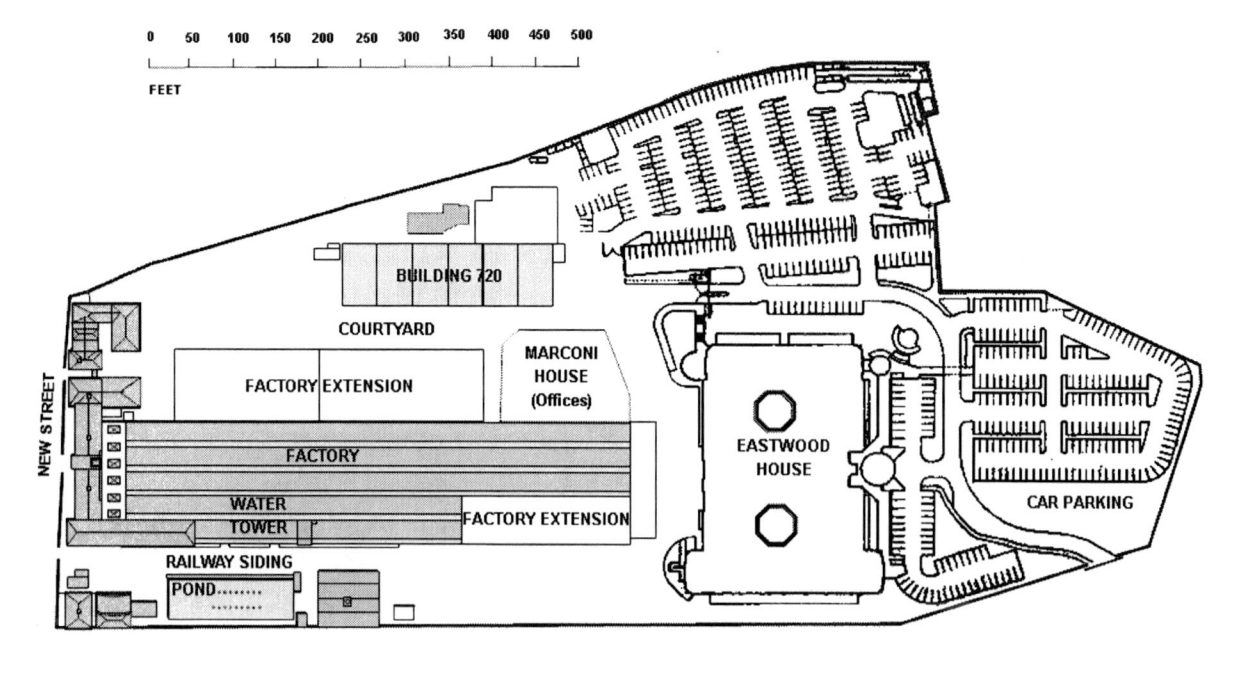

1992-2000

In 1992 the decision was made to remove the numerous buildings, and construct a new building in their place, to be named Eastwood House after Dr Eric Eastwood. Dr Eastwood achieved the rank of Squadron Leader in the RAF during World War II (mentioned in despatches) while working on radar. He was Director of Research at the Marconi Wireless Telegraph Company, 1954-1962, at English Electric, 1962-1968, and at the General Electric-English Electric Companies, 1968-1974. From 1974 he was a Consultant for GEC-Marconi Electronics Ltd and for the GEC-Hirst Research Centre.

Additional land was purchased to be able to extend the Company Car Park. A model of the proposed changes was placed in the foyer of Marconi House. The new building was originally intended for use by Marconi Communication Systems Ltd (MCSL) staff, indeed I remember plans being drawn up for office locations. Marconi Radar Systems Ltd (MRSL) had been planning their own new building at either Writtle Road or Great Baddow when GEC issued a policy statement and simply allocated the new building to MRSL. No debate was offered. The new building though part of the same New Street site was now operated by MRSL and upon its completion in 1994, MRSL moved from Crompton's Works into Eastwood House. Greater confusion occurring in December 1998 when Marconi Radar was transferred to British Aerospace and Alenia Marconi Systems was founded. Eastwood House is now the property of British Aerospace, later to become BAE Systems. The logistics of canteens, security, access, parking and shared resources plagued the respective management teams after both events.

2007

By 2007, Marconi's has been taken over by Ericsson and traded under the name of Telent. The New Street Works just houses what used to be the Military Communications Division of MCSL, now called Selex, a branch of the Italian firm Finmeccannica.

With the exception of Eastwood House which belongs to BAE Systems, the remainder of the original Marconi site was been sold to developers. This Company went into receivership and the site is now derelict and suffering from vandalism and weather damage.

A small part of the original building in New Street has a preservation order, which should survive but the fate of the rest, including the water tower and power generator buildings is still unknown.

It is highly likely that the place where Dame Nellie Melba and Lauritz Melchior made their historic broadcasts will follow in the footsteps of the Marconi Building in the Strand that housed 2LO, and the Writtle site, where 2MT (Two Emma Toc) was located.

APPENDIX 2

THE COMPANY 1937 – 1999

The Company was incorporated in the name - Wireless Telegraph and Signal Company Limited on the 20th July 1897.

Changed to :-
Marconi's Wireless Telegraph Company Limited, 24th March 1900.

Changed to The Marconi Company Limited on 19th August 1963.

Changed to GEC-Marconi Limited on 24th April 1990.

Changed to Marconi Electronic Systems Limited on 4th September 1998.

Changed to BAE SYSTEMS Electronics Limited on 23rd February 2000.

The date of the change on the 24th April 1990 I engineered to coincide with my wife's birthday.

Eric Peachey

Key Dates
in the history of
Marconi and GEC

1937 – 1999

1939 - The Marconi Research Laboratories, Great Baddow are opened.

1939-1945 - During World War II, GEC is a major supplier to the military of electrical and engineering products. Significant contributions to the war effort include the development of the cavity Magnetron for radar, with advances in communications and the mass production of electric lighting.

1946 - English Electric takes over Marconi's Wireless Telegraph Company Limited, and acquires 42% of the shares in the Marconi International Marine Communication Company Ltd., Marconi Instruments Limited and small subsidiaries

1961 - GEC takes over Radio and Allied Industries.

1963 - Marconi's Wireless Telegraph Company becomes The Marconi Company. Arnold Weinstock becomes Managing Director of GEC.

1964 - New premises of The Marconi Company at Waterhouse Lane, Chelmsford are opened.

1967 - GEC acquires Associated Electrical Industries (AEI) in 1967, which encompassed Metropolitan-Vickers, BTH, Edison Swan, Siemens Bros, Hotpoint and W.T. Henley.

1968 - GEC and English Electric amalgamate, incorporating The Marconi Company, Elliott Bros, Ruston and Hornsby, Stephenson, Hawthorn & Vulcan Foundry, Willans and Robinson and Dick Kerr

1974 - GEC acquire Yarrow Shipbuilders.

1979 - GEC acquire Avery.

1988 - GPT created by GEC and Plessey.

1989 - Plessey acquired by GEC and Siemens. Joint investment by GEC and Compagnie General D'Electricitie (CGE), formed the power generation and transport arm, GEC ALSTHOM.

1990 - The movements towards electronics and modern technology, particularly in the defence sector, mean a digression from the domestic market for electrical goods. GEC acquires parts of Ferranti.

1995 - GEC acquires Vickers Shipbuilding and Engineering Ltd (VSEL).

1996 - Lord Weinstock retires. George Simpson becomes Managing Director. Soon there is a major reorganisation and Express Lifts, Satchwell Controls, AB Dick, the Wire and Cables Group, Marconi Instruments and GEC Plessey Semiconductors are sold.

1998 - GEC Head Office moves to One Bruton Street, London. GEC acquires the US defence electronics company TRACOR and becomes part of Marconi North America, a Marconi Electronic Systems Company.

1999 - In January, the announcement of a proposed merger of GEC's Marconi Electronic Systems business with British Aerospace. In April, GEC acquires the US telecommunication network products Company RELTEC, and announces the proposed acquisition of US internet switching equipment company FORE Systems.

On 30th November, GEC is renamed Marconi plc.

BIBLIOGRAPHY

For the reader interested in pursuing more about the career of Marconi there is a huge body of writing and books both in English and Italian that describe Marconi's work and achievements. The following list is by no means fully comprehensive, but covers most aspects.

- Baker, W.J., A History of the Marconi Company, Methuen & Co., London (1970).

- Vyvyan, R.N., Wireless Over Thirty Years, George Routledge & Sons, London (1933). Reprinted as Marconi And Wireless, EP Publishing Limited, Yorkshire, England (1974).

- Bussey, Gordon, Marconi's Atlantic Leap, Marconi Communications 2000, Coventry, England (2000).

- Kemp, George, 'Extracts from the diary of G. S. Kemp.' Marconi Archives, Marconi plc.

- Wander, Tim. '2MT Writtle The Birth of British Broadcasting.' ISBN 978-07552-0607-0. Authors Online - 2010.

- Wander, Tim. 'A Kind of Magic – The Birth of the Wireless Age.' Authors Online - 2012.

- Maria Christina Marconi, 'Marconi My Beloved' ISBN 0-037832-39-1 Dante University of America Press - 2001

- H.E. Hancock. 'Wireless at Sea -The First Fifty Years' Marconi International Marine Company - 1950

- Weightman, Gavin, 'Signor Marconi's Magic Box'   Harper Collins

- Marconi, Degna. 'My Father, Marconi (2nd ed. Rev.)' Ottawa, Canada: Balmuir Publishing Ltd., 1982.

INDEX

Figures in italics refer to illustrations.

ACE Equipment	279
Alum Bay	5, 7
Alexandra Palace	120ff, 125, 268
Anderson Shelter	143
ARC Project	277ff, 281ff, *269*
ARC Works	149, 308
Auxiliary Units	175ff
Baird, John Logie	120ff
Bawdsey Manor	117ff
BBC	97ff, 120ff, 222, 266, 284, 302
Bletchley Park	140
Boer War	12, 36, 68, 138
Bride, Harold	25
Broomfield Station	30, 59, 68. 71, 112, 159
Building 46	188, 196, 203, 220, 262, 265, 342
Building 720	150, 153, 239, 243, 253, 263, 275, 277, 279, 283, 286, 334
Burrows, Arthur	80, 83ff, 86ff, 95
Californian, SS	23ff
Carpathia, SS	24ff
Chain Home (Radar)	68, 70, 105, 117
Chelmsford	7ff, 11ff, 30ff
Clifden Station	13, 14, 66, 92
Cottam, Thomas	24, 26
CR100 Receiver	159, 212, 241
Crompton Works	8, 112, 145, 149, 152, 207, 308
Crookhaven Station	11
Croydon ATC	91, 95ff

Cubitt & Son   34
Cuthbert Hall, E   11, 15

Daily Mail Newspaper   82, 89
Dalston Works   15, *16, 17,* 31
Daventry   106ff, *111*, 116, 266
Direction Finding (D/F)   68ff, 138ff
Ditcham, W.T   79ff
Doodlebug (V1)   144, 162, 168, 184
Dunn William   31

Eaglehurst   43, 75
Eastwood House   131, 150, 219, 281, 311, *352*, 353
Eckersley, Peter Pendleton   96
Elettra (ship)   102, 109
EMI   120ff

Florida, SS   27
Flood-Page, Samuel   11
Fore Systems   306
Franklin, C.S.   76, 104, 115, 122, 142, 187, 281

GEC   198, 255, 269, 271, 302ff
Glace Bay wireless station   13, 66
Grant, Admiral   142, 153, 198, 200
Great Baddow Research   118, 131, 136, 176, 196, 212, 251, 255, 268, 301, 340
Grinnell Fire System   35

Hall Street Works   7ff
Haven Hotel Station   7, 11
Hertz, Heinrich   2ff
Hoffmann Company   80, 112, 145, 146ff, 149ff, 157, 161ff, 184, 194, 212
Home Guard   149, 153, 158, 171ff

Imperial Wireless Scheme   20, 64, 103ff, 107, 338
Isaacs, Godfrey C   18ff, 24, 30ff

Ismay, Bruce	24
Jameson-Davis, Henry	1, 11
Kemp, George	10, 281
Kilostream	295ff
Lodge, Oliver	2, 19, 20
Luftwaffe	182, 184, 308, *343*
Mayo John	305
Maxwell, James Clerk	2
Marconi-Adcock D/F	138
Marconi, Guglielmo	1, 4, 15, 18, 25, 64, 74, 98, 102, 109, 122, 157, 306
Marconi House, The Strand	43, 74, 77, 95, 98
Marconi International Marine Company Ltd	15, 67
Marconigram	14, 22, 98
Marconiphone	100ff
Melba, Dame Nellie	82ff, *86*
Melchior Lauritz	92, *94*
MOFA Project	277
Morrison Shelter	143, 160, 162. 190
MYRIAD Computer	216ff
MZX	76ff
Navy, Royal	5, 7, 9, 12, 22, 65ff, 75ff, 103, 134, 138, 277
Newfoundland	13, 23, 25
New Street	
- Bomb Disposal	147, *148*
- Bombing	149
- Construction	31ff
- Canteen	212, 218, 226ff
- International Radio Telegraphic Conference visit	16ff
- Fire Brigade	131, 161, 229ff, *251*
- Fire Engine	*230*
- Masts	58ff, *61, 62,* 78, 87

- Telephone Exchange   232ff

Northcliffe, Lord   82ff

PCGG   82
Phillips, John George   26
Pochin, E.A.N   12
Poldhu Point Station   9, 11, 13ff, 30, 42ff, 66, 192
Post, Frederick   59
Priddle, E.T   11

R1155 Receiver   138, 179
Radar   109, 114ff
RCA   122
Reltec Systems   306
RENACE   279
Republic, SS   27ff
Round, Captain Henry Joseph   68, 69, 71, 79

Sayer, Winifred   80ff
SELEX   229, 310ff
Simpson George   280, 305ff
Shoenberg, Isaac   120ff
Smith, Edward J. Captain   22, 25
SOE   175ff
Sterling Telephone Company   100
Suitcase Spy Set   175ff
SWB Transmitter   187ff, 208ff

T1154 Transmitter   138
Telefunken Company   19, 20. 75
Television   120ff
Telford, Sir Robert   157, 200, 214, 235, 275
Titanic RMS   20ff

UKMACCS Project   277ff, *289, 290*

V2 Rocket	162, 167ff
V2 Wireless	100
VAX Computers	282ff, 287, *292*, 293ff
Voluntary Interceptor (VI)	139ff
Volturno, SS	28ff

Waterhouse Lane	149, 155, 159, 193, 203, 219, 231, 255, 262, 275, 286, 309
Watson Robert (Architect)	31
Watson-Watt, Robert	99ff
Waunfawr station	34
Weinstock, Arnold	216, 219, 259, 302ff
Wireless World, magazine	16, 107, 240
Writtle (Marconi site)	*94*, 95ff, 112, 138, 162, 187, 272, 276ff, 292
Writtle Road	193, 197, 207, 308

2LO	97ff
2MT	98
5XX	106ff, *111*

Guglielmo Marconi

A man of vision and so profound
He mastered the air waves and produced a sound
Wireless, telegraph and broadcasting too
Were his intention with help from his crew

The British Post Office encouraged his work
As the Italian Government decided to shirk
To Chelmsford he went to establish his name
For the world to appreciate his great claim

That signals could be sent far and wide
Through the ether with no wires beside
His fame began in this quaint market town
Which over the years has gained some renown

An Italian by birth – a non-academic
He conquered physics the start of an epic
Safety at sea was his main concern
To his yacht Elettra he would adjourn

Here his experiments covered communication
To him a debt is owed by every nation
For safety at sea and entertainment too
The outcome of his work with the famous few

He travelled the world in those far off days
To enlighten others of his latest phase
Communication by telegraph was the easiest way
Of sending a message to others the same day

Coherer and Morse Key the basis of his invention
Commanded the Military to seek his attention
The navy too found need of his equipment and advice
To Control ships at sea whatever the price

Problems encountered were soon overcome
As broadcasting by sound radio was begun
In the 1920's the first broadcast was made
By Dame Nellie Melba an Australian from Adelaide

With so much at stake the UK Government took fright
At the power of his inventions they decided to fight
The outcome was to take his wireless station
And therefore broadcasting went to the nation

It is a true fact that the BBC
Was born from Marconi totally
The equipment they took was his own invention
But the BBC never recognise this contention

Now 100 years on a debt to this man we all must owe
As communication, Radar, Satellites and Computers grow
His great foresight is there to see
How everything he did now affects you and me

Peter A.T. Turrall MBE

About the Author – Tim Wander

Raised and educated in Melton Mowbray in Leicestershire, an Honours Degree in Computer Science from Aston University in Birmingham brought him by chance to work at GEC-Marconi Communication Systems' Writtle site, near Chelmsford in Essex. When the site closed in 1987 he moved to New Street.

Tim spent the first 17 years of his career with various arms of the GEC-Marconi Company worldwide, designing, developing and managing radio, telecommunication and control system projects. He left Marconi's in 1999 and spent three more years in senior management within the electronics industry in the City of London. Tim then decided on a career change, finding time to restore a number of classic Jaguar E type cars, he then became a Project Manager for a series of large building projects around the world.

Tim has written many other books, including 'Marconi on the Isle of Wight', produced for the Centenary of the closure of the Alum Bay, Royal Needles Hotel wireless station. A long held interest in early radio sets inherited from his father and a passion for the early days of radio broadcasting led him to write '2MT Writtle - The Birth of British Broadcasting', first published in 1988. After 22 years the second, completely rewritten and much larger edition was published in October 2010.

He has also written several radio plays based on the New Street and Writtle broadcasts. More books are planned and he has recently completed a detailed account of the first five years of Marconi's work in this country and around the world. Called 'A Kind of Magic - The Birth of the Wireless Age', you can keep track of Tim's new and past titles and contact the author via *2MTwrittle.com*.

Tim's hobbies still include a passion for early Jaguar E Type cars, including a period of historic motor racing and all types of shooting, especially with vintage and historic long arms. He has also worked as Chief Mechanic for each of his children as they entered the competitive and at times destructive world of Go-Kart racing.

Married to Judith with three children, Michael, David and Elizabeth, Tim is now a freelance author, lecturer, consultant and project manager. When he is not working or writing he tries to spend as much time as possible wandering around the beautiful mountains of Southern Spain with his five dogs and a battered old Jeep.

Celebrating 100 years........

The Author outside the New Street Works, January 2012